An Introduction to

Surface

Spectroscopy

For Engineers and Scientists

Jorge López Gallardo

and

Miguel Castro Colín

Dedication

This work is dedicated to the hordes of students that –like sacrificial lambs— have suffered my horrible teaching while showing me what works and what doesn't, and to my beloved wife, Rosa, who always complains that I never dedicate any of my books to her.

Jorge A. López
El Paso, Texas
January 23, 2013

To my former students, who were also good teachers, thought they may have not known it. To my teachers and professors, whom I think of when I teach something, even if informally. In the end I guess we are all teachers and learners.
To my daughters and wife.

Miguel Castro Colín
Karlsruhe, Germany
January 3, 2013

Table of Contents

Foreword

One of the most important fields of modern physics is the study of surfaces. Indeed the physical phenomena at the surface are the key of many important industrial applications. Examples of that are catalysis in the oil, chemical and food industry, friction and heat losses in the machinery industry are classical examples. However, as nanotechnology has made its mark on the XXI century new applications have arose such as plasmonics, sensors and optical devices. In a general sense as one of the great inventions of physics, the transistor became industrial in the late quarter of the XX Century.

Electrical engineers are faced with the problem of devices that shrink the size to ~10 Å, then surface phenomena becomes fundamental. For instance the SiO_2 was the most common dielectric material. However, when the size shrank to ~10 Å the dielectric constant was insufficient and researchers launched a search of the whole periodic table until hafnium, a very seldom used material, stepped onto the plate. Similar problems still unresolved are present in silver nanowires where electric conductivity is reduced by surface losses. It is also a subject that is interdisciplinary and cuts trough Physics Chemistry Tribology, Engineering and even Biology. In summary, the surface science is one of the most important areas of modern science.

In this book Jorge Lopez and Miguel Castro Colín present a remarkable vision of surface science from the physicist point of view. This is a very rare and valuable book because most of the available and popular books are written by chemists and present a different point of view. The book of López and Castro Colín is complementary to chemistry-based books and emphasizes the quantum mechanics point of view of the surface phenomena. The presentation on characterization techniques is very unique and insightful. I recommend this book for graduate students of materials, nanotechnology, science, physics and chemistry interested in the subject of surface science.

This work will be a great tool to the education of future scientists.

Miguel José Yacamán
Lutcher Brown Endowed Professor
Chair, Department of Physics & Astronomy
University of Texas at San Antonio

Preface

The world of spectroscopy began its development in the second half of the XX century and has been constantly evolving ever since. From being a set of specialized techniques available only to a few, spectroscopy –in its many facets— has now become a tool chest in demand by engineers and scientists of different backgrounds. Unfortunately, the physics concepts underlying the spectroscopic techniques remain in the field of physics and most users –being from backgrounds other than physics— learn to use commercial spectroscopy devices without ever learning much about their operational principles.

In the University of Texas at El Paso, just like in many other Institutions, students from engineering and science fields come to interact with x ray fluorescence, x ray spectroscopy, Auger electron spectroscopy, to name a few, without ever taking a modern physics course that would teach them, for instance, how x rays are produced, detected or how they interact with atomic electrons. This deficiency, which can be easily remedied with a set of carefully designed lecture notes and exercises, tends to propagate into the realm of applications turning spectroscopy into an obscure subject mastered only by experts; this needs not to be the case.

It is with the goal of providing the minimum physics background needed to understand spectroscopy in its more general terms that we write these lecture notes. They are aimed at an audience of science and engineering students at the senior and graduate level, and mostly composed of chemists, geologists, metallurgists, mechanical and electrical engineers and, yes, physics students, all interested in studying materials through the use of spectroscopic tools.

Assuming only a background of basic classical mechanics, electricity and magnetism, and thermodynamics, these notes focus on explaining how radiation (particles and electromagnetic) interact with matter, and how this is taken advantage of to study materials through the use of spectrometers. Because of the narrowly defined audience, the book is equally limited in scope, most physics processes will be presented more from a phenomenological point of view than from a first-principles fully-theoretical approach. Students ready for more profound treatises will be directed throughout the book to other more complete sources.

Designed as a teaching textbook, a large number of exercises and problems have been included to illustrate concepts and applications. Instructors are encouraged to contact the authors to obtain a complimentary file with the solutions, a test bank and *Powerpoint* files of the chapters expanded with instructive animations and progressive presentation of the examples for in-class use.

Jorge López Gallardo   **Miguel Castro Colín**
El Paso, Texas, USA   **Karlsruhe, Germany**

Introduction

Spectroscopy, the study of the interaction between matter and radiation, has become the tool most commonly used to study materials. Evolving from the prism dispersion of visible light, spectroscopy nowadays is based on the interaction of beams irradiated on materials. Current beams are now composed of electromagnetic radiation or particles and give rise to a plethora of different spectroscopic techniques, such as Auger electron spectroscopy, x-ray energy dispersive spectroscopy, electron spectroscopy for chemical analysis, Rutherford backscattering, and many others.

The first step in studying modern spectroscopy is to understand how radiation interacts with matter and, for this, a review of the properties of electromagnetic radiation and of particles is in order. Chapter 1 starts by reviewing the main properties of electromagnetic radiation (in its different frequencies), particles (such as electrons, protons, neutrons and ions), and atoms and the structures they form. This initial chapter is also utilized to introduce the mass and energy scales relevant for the spectroscopic processes.

Chapter 2 uses the radiation and particle information of Chapter 1 to study the basic processes occurring between electromagnetic radiation and matter. X ray production, x ray fluorescence, photoelectric effect, Compton scattering, and Bragg diffractions are some of the effects covered.

Chapters 3, 4 and 5 use the effects studied in Chapter 2 to see how they are implemented in the study of materials. Respectively, the three chapters focus on x ray fluorescence, x ray photoelectron spectroscopy and Auger electron spectroscopy.

Additional material related to vacuum techniques as well as binding energy tables and a comparison of techniques are included in the appendices. The book closes with an annotated bibliography and references cited.

Chapter 1: Radiation and Particles

This chapter summarizes the basic properties of electromagnetic radiation (both classically and quantum), particles (protons, neutrons and electrons), atoms and solids.

Classical Electromagnetic Radiation

Light, in the classical sense, is a wave of electric and magnetic fields that oscillate perpendicularly to one another but also perpendicular to the direction of propagation. Similar waves oscillating at different frequencies exist although not visible by the human eye. In general terms electromagnetic (EM) radiation is produced whenever electric charges accelerate or decelerate.

For instance, the figure shows a circuit similar to the one used by Heinrich Hertz in 1887 to produce sparks regularly in time, which in turn produce travelling EM waves with the same frequency as the sparks.

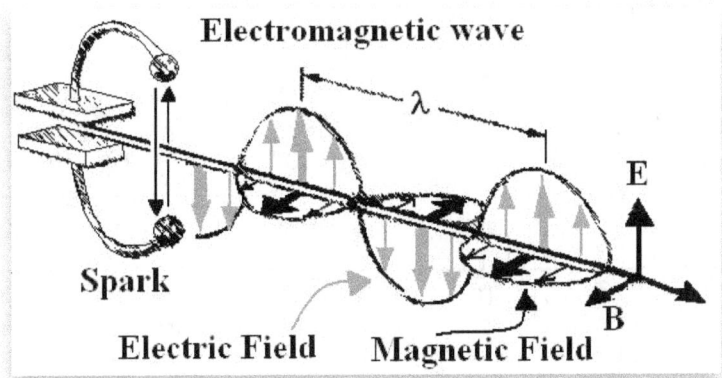

The following figure shows the oscillatory path followed by a beam of electrons inside a magnetron to produce EM waves in a microwave oven. As the electrons move in semi-circles, their acceleration produces EM waves with frequencies of about 10^9 oscillations per second.

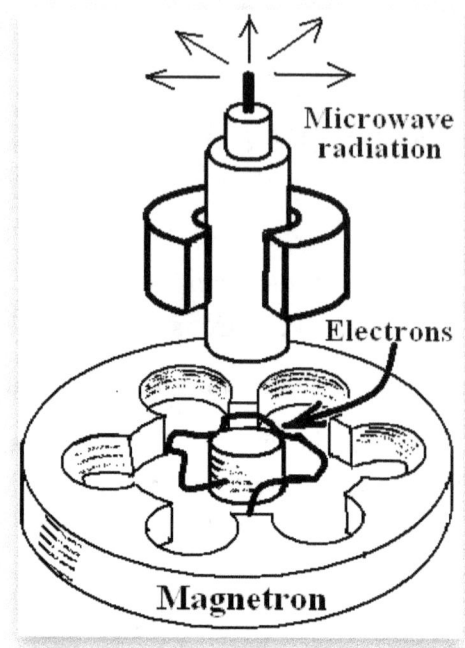

In vacuum light always travels at the speed $c = 2.99 \times 10^8$ m/s, and, as any other wave, its frequency (f), speed (c) and wavelength (λ) are related by means of $c = \lambda f$. The units for the frequency are Hertz (oscillations per second, 1/s) and meters for λ.

Other manifestations of electromagnetic radiation are presented in the following figure, which also shows the nomenclature used for waves of different frequencies and wavelengths. For a more complete reference of nomenclatures as well as of fundamental physical constants consult, e.g., the Reference on Constants, Units and Uncertainty of the National Institute of Standards and Technology (NIST).

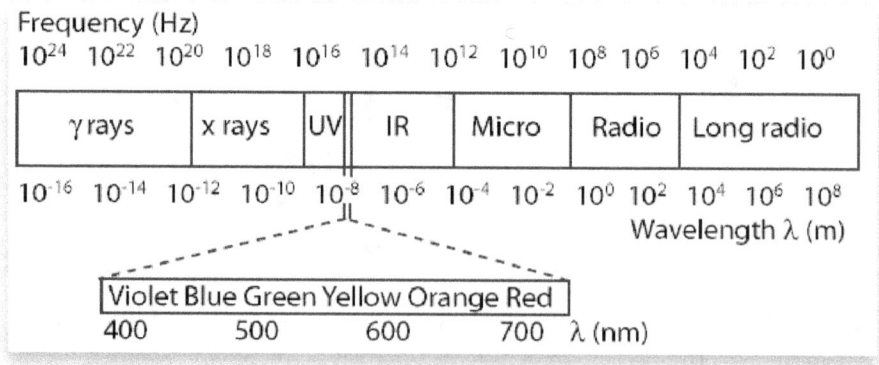

Exercise 1.1

Why can you see through a microwave oven door but the microwave radiation cannot escape? Take the frequency of operation as 2,450 MHz.

Solution

The microwave radiation has a "size"

$$\lambda = c/f = (3 \times 10^8 \text{ m/s})/(2450 \times 10^6 \text{ 1/s}) = 0.122 \text{ m}.$$

The visible light has a "size" of up to 700×10^{-9} m.

The size of the holes of the metal screen of the door is of about 1 mm, i.e. the holes are about 122 times smaller than the microwave radiation (122 mm vs. 1 mm), but about 1428 times larger than the 700 nm visible light (0.0007 mm vs. 1 mm).

Photons

At the microscopic level light can be taken as composed of particle like bundles of electromagnetic fields known as photons. This interpretation was originated to reconcile the wave phenomena exhibited by light, such as diffraction, etc., and particle-like phenomena such as the photoelectric effect.

Although it is impossible to visualize photons, a useful depiction is that of a wave packet containing the crossed EM fields as shown in the accompanying figure. Notice that the wave is now finite in the direction of propagation and that the fields have zero amplitude before and after the packet; i.e. the photons have a finite size.

As in the classical case, the photons are made of oscillatory EM waves and carry energy and momentum in the form of EM fields.

The energy of a photon is related to its frequency through $E = hf$, where E is the energy in Joules, f the frequency in Hz and h is Planck's constant $h = 6.62 \times 10^{-34}$ J s (1 Joule = m^2 kg/s). Although they are massless, photons carry momentum with them given by $p = h/\lambda$, in units of m kg/s.

Unfortunately, the size of a photon (volume, radius, length, etc.) cannot be measured or even defined. For our purposes we can argue that a photon has a size comparable to the size of its source, which in turn is comparable to its wavelength. Thus a gamma ray produced by the motion of nuclear charges, will have a wavelength (and a "photon size") comparable to the nuclear size, i.e. $\lambda \sim 10^{-15}$ m, and an x-ray produced by electrons in an atom will have a wavelength and a photon size comparable to the atomic size, i.e. $\lambda \sim$ Å $= 10^{-10}$ m. Conversely, gamma rays will go through atoms and will only be stopped by nuclei, while x rays will go through, say tissue cells but will be stopped by a wall of atoms, such as that of bone.

Useful Formulae and constants

Photons: $E = h\,f$, $p = h/\lambda$

Rest mass energy $E = mc^2$

Units: 1 fm $= 1 \times 10^{-15}$ m; Å $= 1 \times 10^{-10}$ m; 1 nm $= 1 \times 10^{-9}$ m; 1 μm $= 1 \times 10^{-6}$ m; 1 pm $= 1 \times 10^{-12}$ m; 1 J $= 6.24 \times 10^{18}$ eV; 1 eV $= 1.6 \times 10^{-19}$ J; h $= 6.62 \times 10^{-34}$ Joules s; h $= 4.135 \times 10^{-15}$ eV s; hc $= 1240$ eV nm

Spectroscopy energy unit:
$$cm^{-1} = 1\ cm^{-1} \times (4.135 \times 10^{-15}\ eVs) \times (2.99 \times 10^{10}\ cm/s)$$
$$= 1.236 \times 10^{-4}\ eV$$

Exercise 1.2

A) Estimate the energy of the photons produced in the microwave oven of Exercise 1.1. Again, take the frequency of operation as 2,450 MHz.

B) If the magnetron power output is of 600 W, how many photons are emitted per second?

Solution:

A) $E = hf = (6.62\times10^{-34}$ J s$)(2450\times10^6$ 1/s$) = 16219\times10^{-28}$ J

$= 16219\times10^{-28}$ J $\times$ 6.24×10^{18} eV/J $= 1.01\times10^{-5}$ eV

B) P = energy/time = E/t $\Rightarrow$ energy emitted each second is

$$E = Pt = 600 \text{ J/s} \times 1 \text{ s} = 600 \text{ J}.$$

This corresponds to the following number of photons:

$$N = 600 \text{ J}/16219\times10^{-28} \text{ J} = 3.7\times10^{26}.$$

Protons, neutrons, and electrons

Atoms are composed of three types of particles: protons, neutrons and electrons. The table shows the masses, charges and approximate sizes of these particles; the charges are in units of the charge of an electron, which is 1.6×10^{-19} Coulombs. Notice that protons and neutrons have similar masses, and they are about 2000 times more massive than the electrons; 99.94% of the mass of an atom is concentrated in the nucleus. The masses are also listed in atomic mass units, "u", which represents 1/12 of the atomic weight of a C atom, u = 1.660538921 $\times10^{-27}$ kg.

	Mass	Charge (e)	Size
Proton	1.672×10^{-27} kg 1.00727 u	+1	0.877×10^{-15} m
Neutron	1.674×10^{-27} kg 1.008664 u	0	Same as proton
Electron	9.109×10^{-31} kg 0.000548 u	-1	0, pointlike

Just like photons can behave as particles (see e.g. the section "Photo-electric effect" in the next chapter), particles -such as protons, neutrons and electrons— can behave as waves, i.e. particles also show interference and diffraction phenomena like waves. This property of matter was predicted by Louis de Broglie in 1924 and corroborated experimentally by Davisson and Germer who found diffraction phenomena with particles in 1927.

5

Although to date it is not known what exactly oscillates in a matter wave, it is known that particles have a wavelength given by $\lambda = h/p$, where p is the particle momentum, and beams of particles can produce diffraction and interference patterns when going through atomic layers. More on this topic will be covered in the following chapter.

Another property of particles is that their mass (m) and energy begin to be interchangeable according to Einstein's relation $E=mc^2$. Sometimes part of the mass of a particle becomes binding energy (see, e.g. Exercise 1.5), while sometimes it completely disappears to produce, e.g., high energy photons (see Exercise 1.4).

Yet another property of the particles is an intrinsic magnetic field they possess; as this field appears to be produced by a spin of the particle (and its internal charges) on its own axis, it carries the apt name _spin_. When particles are in the presence of an external magnetic field, they rotate to align their internal field with the external one; the fact that this alignment can only occurs at some discrete angles, makes this phenomenon different in nature to the classical case in which the angle can take up any value. Atomic electrons in an atom, for instance, tend to arrange themselves in anti-aligned pairs to minimize the interaction energy. The figure shows a hydrogen atom with the only two possible alignments of the electron in the proton field: aligned (top) and anti-aligned (bottom); given the difference in energy, electrons in the top state tend to switch to the bottom

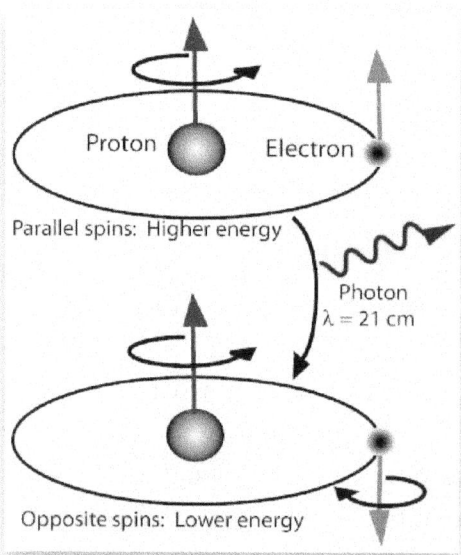

Proton   Electron

Parallel spins: Higher energy

Photon
$\lambda = 21$ cm

Opposite spins: Lower energy

state emitting the excess radiation in an EM photon of 21 cm of wavelength; in outer space this occurs in an average time of 7 years and forms the basis of radio astronomy.

Exercise 1.4

Positron-electron tomography is an imaging technique used in medicine to produce three-dimensional pictures of internal body parts. It works by ingesting a radioactive substance (e.g. F^{18}, C^{11}, N^{13} or O^{15}) that emits *positrons* (positive electrons, properly speaking *antielectrons*) which, upon encountering an electron, get annihilated along with the electron producing a flash of two gamma rays. The γ rays can then be used to produce a map of the emitting internal organ. What is the energy (in eV) of the produced gamma rays?

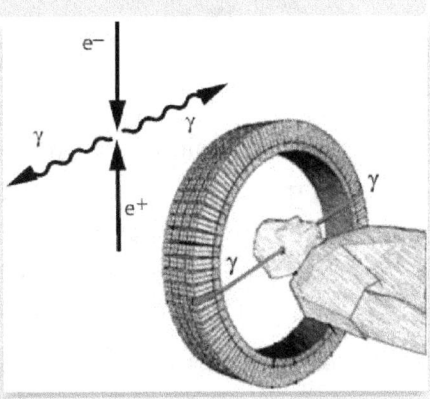

Solution:

Since the e^+ and the e^- are being annihilated, their whole mass turns into the gamma rays energy.

$$E_\gamma = m_e c^2 = 9.109 \times 10^{-31} \text{ kg} \times (2.99 \times 10^8 \text{ m/s})^2 = 8.14 \times 10^{-14} \text{ J}$$

$$= 8.14 \times 10^{-14} \text{ J} \times 6.24 \times 10^{18} \text{ eV/J} = 0.508156 \text{ MeV}.$$

Using more decimals in m_e, c, etc. yields the exact answer of 0.511 MeV.

Exercise 1.5

The "spin flip" that occurs between the parallel spin state of the H atom and the antiparallel spin state is analogous to the effect used in the medical imaging technique known as nuclear magnetic resonance (NMR) induced by high frequency EM radiation.
A) Calculate the difference in energy between the two H states.
B) What type of radiation could induce the reverse spin flip?

Solution:

A) Since the energy difference produces a 21 cm photon, such energy equals

$$E = hf = hc/\lambda = 1240 \text{ eV nm}/21 \times 10^{-2} \text{ m} = 5.90 \times 10^{-6} \text{ eV}$$

B) Radiation of the same energy, i.e. radio waves.

Atoms

The atom is the basic unit of matter and consists of a dense central nucleus surrounded by a cloud of electrons. The nucleus is composed of protons and neutrons bound through the nuclear (strong) force. The protons and electrons interact through the electric force. The nuclear force has a range of about 5 fm (1 fermi or femtometer is 1×10^{-15} m), and the electric force is of infinite range.

The binding energy of neutrons and protons in the nucleus is of the order of 8 MeV (1 eV=1.6×10^{-13} Joules), and it equals the minimum amount of work needed to separate a nucleon from the nucleus. Electrons in atoms have binding energies that vary from hundreds of eV for core electrons (lowest energy levels), to tens of eV for valence electrons in metals and semiconductors, to an eV or a fraction of it for conduction electrons in metals. In all cases the nuclear binding energy is thousands of times larger than the binding energy of an electron.

As illustrated in the figure, atoms are of size of the order of angstroms (Å) while nuclei are of the order of up to 15 fermis, i.e. the atom is between 20,000 and 145,000 times larger than the nucleus.

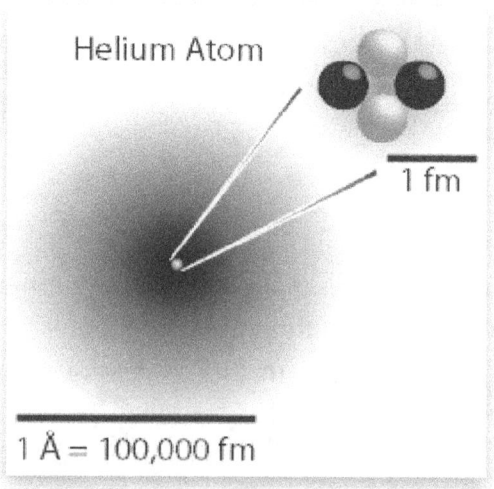

Helium Atom

1 fm

1 Å = 100,000 fm

The table shows the ranges for atomic masses, sizes, etc. and the periodic table in Appendix 2 (Table A2.4) shows the atomic radii as determined by Slater in 1964 according to the distance between atoms in covalent bonds; the radii of inert gases (and other elements) cannot be determined this way.

Atomic scales	
Mass range	1.67×10^{-27} kg to 4.52×10^{-25} kg
Size (diameter)	62 pm (He) to 520 pm (Cs)
Protons	1 (H) to 109 (Mt)
Atomic mass	1 (H) to 266 (Mt)

Atoms tend to have equal number of protons and electrons and are electrically neutral but can lose or gain electrons and become electrically charged "ions". The chemical properties of an atom are due mainly to the structure of the electrons in the atoms, but the physical properties are due to the number of protons, neutrons and electrons they contain.

The number of protons determines the *atomic number*, the number of neutrons determines the *isotope* of the element. The total number of protons, neutrons and electrons determine the *atomic weight*, which is not a weight but a mass expressed as a fraction of the mass of a carbon atom.

For instance, the figure shows three nuclei with one proton, hence the atomic number of all three is 1. But since the number of neutrons varies the three nuclei correspond to three isotopes. Although the atomic weights

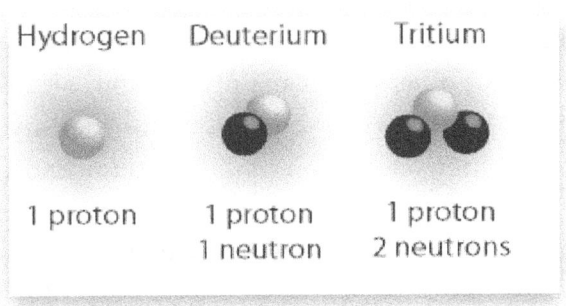

9

should be 1, 2 and 3, they are in fact 1.00794 u, 2.014102 u, and 3.0160492 u; with "u" representing the atomic mass unit equivalent to 1/12 of the atomic weight of a C atom, $u=1.660538921 \times 10^{-27}$ kg.

Exercise 1.6

A) Use the masses of protons, neutrons and electrons listed before to estimate the mass of a carbon atom which has six protons, six neutrons and six electrons.

B) If the actual mass of a carbon atom is 12.000 u, how do you explain the difference between your estimation and the actual value? Can you estimate the average nucleon binding energy from this difference? (Hint: think of mass becoming energy through Einstein's relation $E = mc^2$)

Solution:

A) $M = 6 \times (1.00727 + 1.008664 + 0.000548)$ u $= 12.098892$ u

B) The mass difference, 0.098892 u, turned into binding energy according to

$E = mc^2 = 1.477926134779788 \times 10^{-11}$ J

$\quad = 1.477926134779788 \times 10^{-11}$ J $\times 6.24 \times 10^{18}$ eV/J $= 92.2$ MeV.

Neglecting the electronic binding energy, the average binding energy per nucleon would be BE $= 92.2$ MeV/12 $= 7.68$ MeV, close to the expected value of 8 MeV.

Solids

Because of their electronic structure, the atoms have the ability of attracting other atoms mainly through electromagnetic interactions and, depending on conditions of density, temperature, atomic composition, etc., form rigid structures known as solids. In addition to the usual physical properties (volume, density, etc.) solids can be characterized by the internal structure of the atoms. The atoms in a

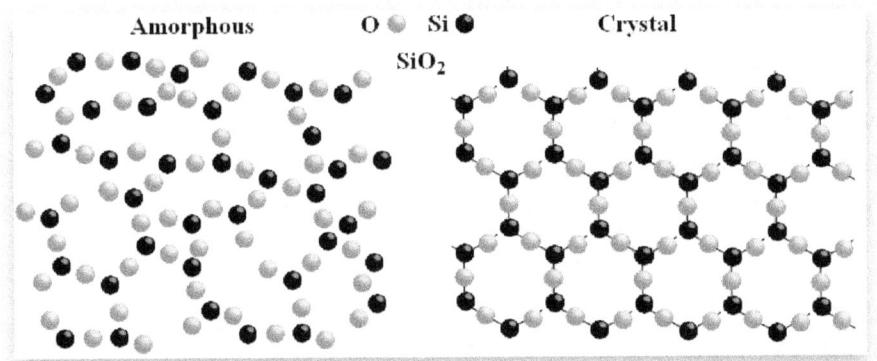

solid are bound to each other, either in a regular geometric lattice or irregularly.

Examples of regular patterns are crystals, metals and even ice, and examples of amorphous solids are polymers and glasses. The figure shows schematically one possible layer of amorphous quartz (left panel) and of crystalline quartz (right panel).

Atoms are held to one another by means of electric forces. When one atom has an excess of positive charge and attracts another atom that has an excess of negative charge it is said that the bond is *ionic*. When two neutral atoms bind to one another sharing electrons, the binding is said to be *covalent*. In the case of metals the bonds occur between positively charged metal ions and atoms with free conduction electrons; this is known as *metallic* bond. In terms of strength, covalent bonds are stronger than ionic bonds which are stronger than metallic bonds. But all bonds are of the order of a few eV, smaller than the binding energies of atomic electrons and much smaller than the nucleon binding energies.

The distance between atoms in a material is of the order of a few angstroms, equivalent to a few atom sizes. Although the atoms in solids are fixed, they can vibrate and rotate in place; under some conditions atoms can migrate. Usual values of the energy of vibration and rotations are in the tenths of eV for vibrations and in the 10^{-5} eV for rotation.

Exercise 1.7

EM radiation of various frequencies is directed to a crystalline solid as shown. Explain what happens to radiation of the following types when they hit the surface: A) radio waves, B) x-rays, C) γ rays

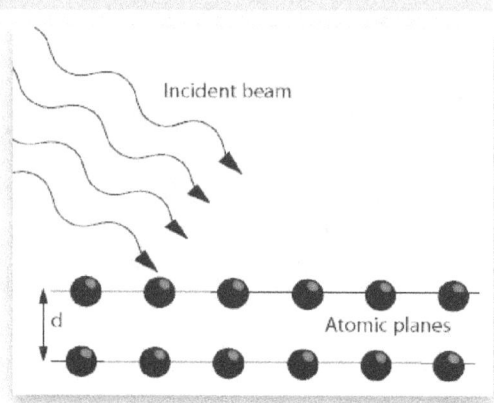

Solution:

If the EM fields in the wave hit an atom, they will make the atomic charges oscillate and re-radiate with the same frequency. However, radio waves being of λ of the order of meters will hit zillions of atoms at different times making them radiate incoherently producing specular reflection. X-rays being of sizes comparable to the interatomic spacing will make the atoms vibrate close to being in phase producing Bragg diffraction, i.e. a strong reflection at a given angle (see next chapter). Gamma rays being of λ thousands of times smaller than the interatomic spacing will simply miss the atoms and will go through the material, probability indicates that a (very) few will hit a nucleus.

Exercise 1.8

Explain which type of EM radiation can produce the following effects:

A) Rotation of atoms in solids

B) Vibration of atoms in solids

C) Emission of an atom from a solid

D) Emission of an electron from an atom

E) Emission of a nucleon from a nucleus

Solution:

A) Rotation needs μeV thus microwaves will do it (the food is warmed in a microwave oven by rotation of the water molecules in the food.

B) Vibrations need tenths of eV, thus infrared (IR) light will do it. In fact the vibrations of our own body molecules produce IR radiation, that is how night vision goggles operate.

C) Atomic bonds need a few eV to be broken, thus UV will do it. UV radiation is used for food processing (to break bonds of viruses) and mattress cleaning, among other uses.

D) Conduction electron emission can happen with energies of eV and higher, thus visible light and higher energy photons can do it through a process known as photoelectric emission; solar cells operate in the visible.

E) To emit a nucleon energies of MeV are needed, gamma rays are the only ones that can do it.

Further reading

Being this a general topic, any book on "Modern Physics" (which is not so modern anymore) will serve the purpose. Textbooks at this level that the authors have used in the past are Beiser, Krane, Serway and Tipler; see the Bibliography for the proper reference. Extensive material can also be found online, in particular in Wikipedia, and in the Hyperphysics site.

Problems

Problem 1.1

What should be the minimum size of a satellite dish if we had interest in receiving a signal from the satellite Intelsat-4 (IS-4)? Although relevant, it should not be considered the specific geographical location of the dish. Its lowest frequency of emission was about 3,704 MHz.

Problem 1.2

A) Estimate the energy of the photons produced by the satellite IS-4 in Problem 1.1. Consider the same frequency as in the problem 3,704 MHz.

B) If the power emission of digital TV requires about 300 W at a frequency of 470 MHz, how many photons are emitted per second?

Problem 1.3

A) Estimate the mass of carbon-14 (^{14}C), knowing that it has six protons, eight neutrons and six electrons. This type of carbon is a radioactive isotope of carbon. Note that it has six electrons.

B) The tabulated mass of ^{14}C is 14.003241 u. What is then the nucleon binding energy considering the result obtained in A)? (Hint: bear in mind Einstein's equivalence relation between energy and rest mass, $E=mc^2$)

Problem 1.4

Thermal neutrons have energy of about 25 meV, what is their wavelength? Can these neutrons be used to investigate the structure of materials? Please explain. Recall that neutrons are massive particles but nevertheless can be associated to wave-like phenomena.

Problem 1.5

When observed from a distance, two objects close to each other appear as a single one. The *resolution* of an instrument (lens, antenna, etc.) is the minimum distance at which the instrument can separate images expressed as an angle. In general, the resolution (in radians) is proportional to the wavelength of the EM radiation being received and inversely proportional to the physical size (diameter) of the instrument, i.e. $r \propto \lambda/D$. With this in mind:

A) What can be done to increase the resolution of an instrument?

B) In Socorro, New Mexico, there is an array of radio telescopes and the distance between the radio antennas that form the array can be varied. Why do you think the radio astronomers may want to do that?

Problem 1.6

H.G. Wells published in 1897 a novel named "The invisible man", which prompts the following interesting question. If an invisible woman could exist,

A) what properties the stuff she would be made of should have?

B) Would she be able to see visible light?

C) What type of "eyes" she would have to have to see radio waves, UV, IR, x rays?

Chapter 2: Basic Processes

This chapter will review several physical processes that involve the EM radiation, particles, atoms and solids introduced in Chapter 1. These processes involve the production of electron beams, x rays, spontaneous and stimulated emission of photons, etc.

Thermionic evaporation of electrons

Many applications require the use of an electron beam, such as x ray production, Auger spectroscopy, and even cathode ray tube (conventional) TV sets, to name a few; in this section we will review the process of producing an electron beam.

Most beams start by obtaining electrons from a metal through the process of thermionic emission in which a metal is heated to a high temperature and an electric field is used to strip electrons from the metal. The chart shows the ionization energy for elements in their ground state; metals have the lowest ionization energies while inert gases have the largest.

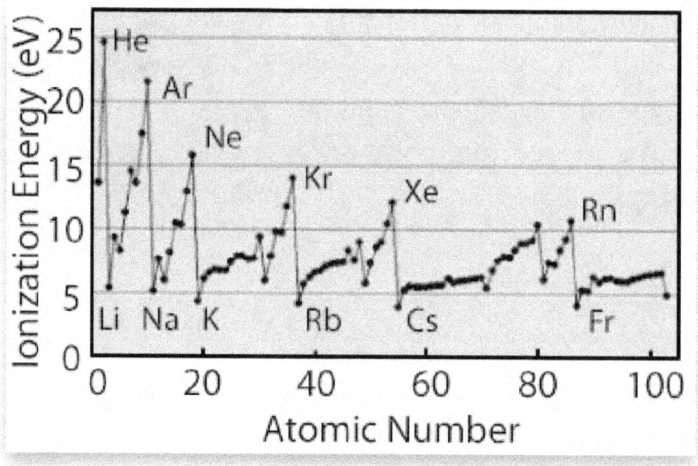

To facilitate the emission of electrons, metals are heated for the electrons to populate higher energy levels where the binding energy is of the order of an eV and smaller. At the same time electric fields are introduced to lower even more the energy barrier through the so-called Schottky effect, which results in the liberation of electrons.

Once the electrons have been freed, they can be moved, focused and accelerated by means of electric fields. Operating temperatures

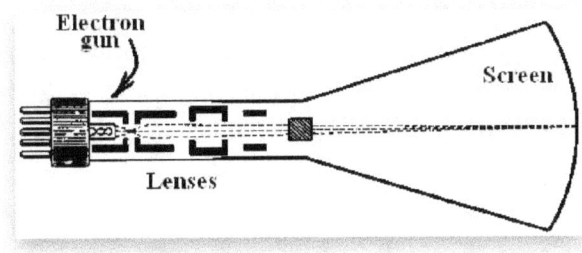

are of the order of 1,000 K or higher, and Schottky fields are of hundreds of millions of volts/m. The figure shows a common electron gun of the type used in conventional TV screens.

Exercise 2.1

In the Sun atmosphere helium (He) is found ionized, i.e. instead of having two electrons it only has one. This is the result of collisions between free electrons in the atmosphere with atomic electrons in the He atoms. What temperature should the free electrons have to be able to ionize He?

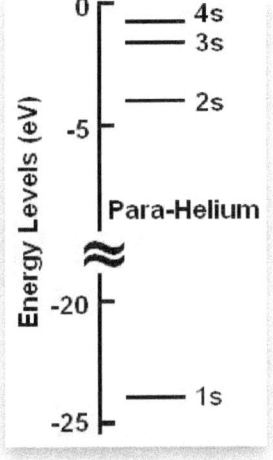

Solution:

He is presumed to have one electron in the 1s state and the second one either with the spin antiparallel to the 1s electron forming an $S = 0 = (1/2) + (-1/2)$ parahelium atom, or in parallel to the 1s electron forming an $S = 1$ $= (1/2) + (1/2)$ orthohelium atom. Assuming the former and using the data from the figure, the ionization would require about 4 eV of energy, and the culprit free electron would have to have at least such amount of energy. Thus the temperature of the electron gas should be $E = (3/2)kT \Rightarrow$

$$T = 2E/3k = 2 \times 4 \text{ eV} / 3 \times 8.617 \times 10^{-5} \text{ eV/K} = 30,946 \text{ K}$$

Production of x rays

EM radiation is always produced whenever electric charges are accelerated or decelerated. In the case of x rays, these are produced by a beam of electrons which is stopped by a metallic target; this process is known as *bremsstrahlung*, which in German literally means braking radiation. The figure shows a diagram of an x ray tube.

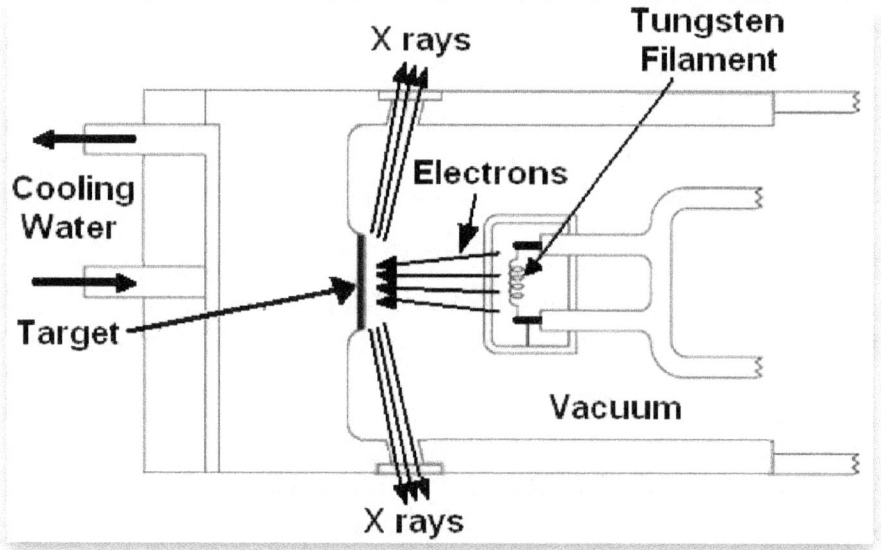

Such stopping process is extremely complex and does not result in the production of x rays with a unique frequency as, e.g., the radio waves produced by an antenna; in spite of the complexity, the x rays produced can be grouped in two categories. The most abundant group is generated directly by the bremsstrahlung, i.e. the slowing down of electrons by the target, which results in fast, medium and slow decelerations that produce x rays with a continuous distribution of energies (or wavelengths).

If one operates an x ray generator for some amount of time, captures the radiation produced and identifies their energies, one could plot the number of photons captured (intensity) versus their energy. An example is presented in the accompanying figure obtained by irradiation of an electron beam of 100 keV on a tungsten target; the bulk of the curve is produced by the bremsstrahlung radiation.

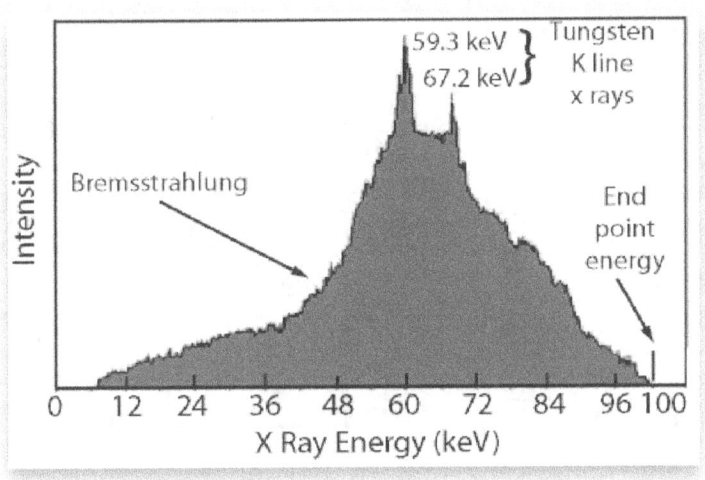

The second component of the produced x rays is the radiation produced by excitation and de-excitation of the atomic electrons in the target itself. Synoptically, the incident electron kicks a core electron out of the target atom and another electron of a higher level jumps to occupy its place emitting an x-ray photon. Usually, for the beam energies used, the emitted electrons are from the K, L or M shells, and the energy of the photon emitted equals the difference between these levels. The nomenclature used for the x ray spectroscopy is $K_{\alpha 1}$, $K_{\alpha 2}$, … for all x rays produced by electron transitions from the L shell into the K shell, $K_{\beta 1}$, $K_{\beta 2}$, … . for transitions from the M shell to the K shell, $L_{\alpha 1}$, $L_{\alpha 2}$, … for transitions from the M shell to the L shell, etc. When all lines $K_{\alpha 1}$, $K_{\alpha 2}$, etc. are very close in energy they are denoted generically simply by K_{α}.

The previous spectrum shows two such transitions prominently: K_α at 59.3 and K_β at 67.2 keV, but certainly other transitions (e.g. L_α, L_β, etc.) must also be present but are not as pronounced as the K_α, K_β. The intensity (emissions per second) of x rays of different energies depends on several factors including the energy distribution of the electron beam, angle of incidence of electrons, angle of capture of x rays, etc.

Other methods of production of x rays use cyclotrons, synchrotrons and pyroelectric crystals.

Exercise 2.2

Use the electron energy levels of tungsten to explain the energies of the two x ray peaks shown in the previous figure.

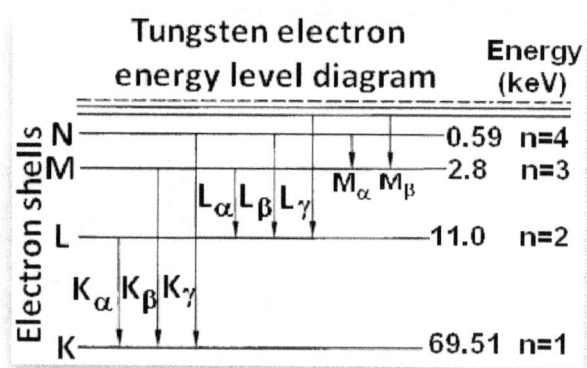

Solution

The diagram indicates the possible transitions that produce x rays. For instance, the energy of a K_α x ray produced by the decay of an electron from an energy level in the L shell into a level in the K shell is of the order of

$$69.51 - 11.0 = 58.51 \text{ keV},$$

and that of a K_β transition is

$$69.51 - 2.8 = 66.71 \text{ keV}.$$

Both quantities are in close agreement with the energies of the peaks shown in the previous figure.

Photoelectric effect

Under some circumstances, when light shines on metal, electrons are emitted by the metal, this is known as the photoelectric effect and its explanation won Einstein his Nobel Prize in 1921. In a nutshell, the photoelectric effect is simply the collision of a photon and an atomic electron resulting in the emission of the electron. In general this effect can also occur in materials other than metals.

The energetics of the process is equally simple, the photon must have enough energy as to break the electron bond and send it out with some kinetic energy. The minimum amount of energy that the photon must have for this process to occur is called the _work function_ and it equals the binding energy of the emitted electron plus a little bit more to overcome the attraction of the metallic surface (now positively charged after losing an electron). Photons with energies larger than the work function will endow the electrons with a larger kinetic energy. The table shows work function values for several metals.

Work functions (eV)											
Al	4.08	C	4.81	Au	5.1	Hg	4.5	Pt	6.35	U	3.6
Be	5.0	Cs	2.1	Fe	4.5	Ni	5.01	Se	5.11	Zn	4.3
Cd	4.07	Co	5.0	Pb	4.14	Nb	4.3	Ag	4.73		
Ca	2.9	Cu	4.7	Mg	3.68	K	2.3	Na	2.28		

Several points of interest.
- Increasing the intensity of photons (i.e. number of photons impinging on the surface per unit time) will result in an increment of the number of electrons emitted, but not on the kinetic energy of the electrons.
- Increasing the energy of the photons will result in an increment of the kinetic energy of the electrons, but not on the number of electrons emitted.
- For every material there will be a minimum energy for the process to take place, photons with energies smaller than the work function will not produce the effect.

21

- The maximum kinetic energy of the electrons is related to the photon energy and the work function through KE = hf − Φ, where KE is the kinetic energy, h is Planck's constant, f is the frequency of the incident light and Φ stands for the work function. Notice that electrons can lose energy as they travel through the material towards the surface, thus KE is the maximum value of the kinetic energy that electrons can have; if the kinetic energy is zero, then hf = Φ.

Exercise 2.3

Millikan studied the photoelectric effect in 1916 obtaining the data shown on the chart. The abscissa shows the frequency of the photon used, and the ordinate the

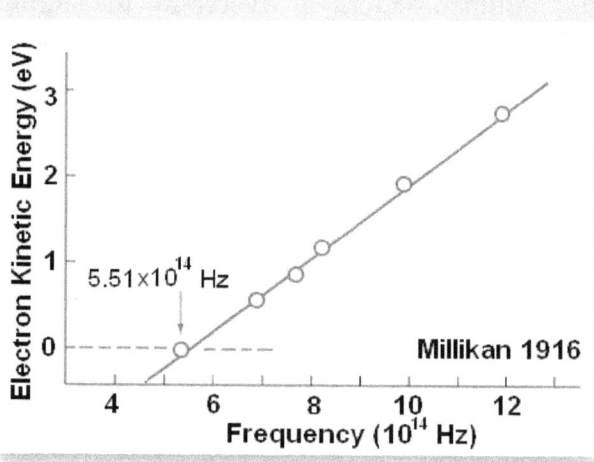

kinetic energy of the emitted electron. Based on this information:

A) Determine the material used by Millikan.

B) Can the effect be produced with far-infrared light?

C) Use the data to estimate Planck's constant.

Solution

A) Since the effect does not occur for photons with frequencies smaller than $f_0 = 4.3 \times 10^{14}$ Hz, that implies that the work function should be equal to

$$\Phi = hf_0 = 4.135 \times 10^{-15} \text{ eV s} \times 5.51 \times 10^{14} \text{ Hz} = 2.278 \text{ eV}$$

consequently the material must have been sodium.

B) Far IR corresponds to frequencies of 10^{12} Hz, too small to produce the effect.

C) Since KE = hf − Φ, the slope of the line KE vs. f is Planck's constant h, consequently

$$h = \Delta KE/\Delta f \approx 2.7 \text{ eV}/(12\text{-}5.51) \text{ Hz} = 4.16 \times 10^{-15} \text{ eV s},$$

not too distant from the actual value of $h = 4.135 \times 10^{-15}$ eV s.

Compton and Thomson scattering

Most collisions between x rays and bound electrons do not result in the absorption of the photon as in the photoelectric effect. It is far more probable for collisions to occur between photons and electrons in the outer orbits and, in some cases, the x ray photons will not be absorbed but will scatter from the electron transferring some or none of its energy. Cases with energy transfer (i.e. inelastic scatterings) are called Compton scattering and cases with no energy

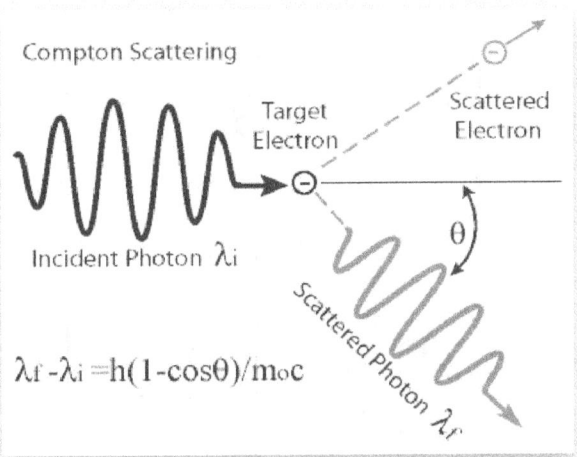

transfer (elastic scatterings) are called Thomson (or Rayleigh) scatterings. In general, scattering can be used to investigate the electronic density distribution of a material, as well as for imaging, radiobiology, and radiation therapy among others; in particular x ray Compton backscattering is nowadays used in imaging technology in airport security devices.

Compton scattering is named in honor of Arthur Compton who discovered it in 1923. In this process, as shown in the figure, the incident photon loses energy and ends with a longer wavelength while freeing an electron. Mathematically, conservation of energy and momentum yield the relationship between the final x ray wavelength (λ_{final}) and the initial one ($\lambda_{initial}$):

$$\lambda_{final} = \lambda_{initial} + h(1-\cos\theta)/m_o c$$

where θ is the angle at which the photon is scattered, m_o is the electron mass and c is the speed of light, notice that $h/m_o c = 0.02426$ Å, quantity commonly denoted by λ_C, the Compton wavelength.

In his original experiment, Compton scattered Mo K_α x rays of $\lambda=0.07078$ nm off atomic electrons in a C target and captured them at different angles measuring the wavelength. His data, (see figure) at 90° shows the original elastic scattering of the Mo x rays (at 0.07078 nm) and the Compton scattered x rays at $\lambda = 0.07314$ nm.

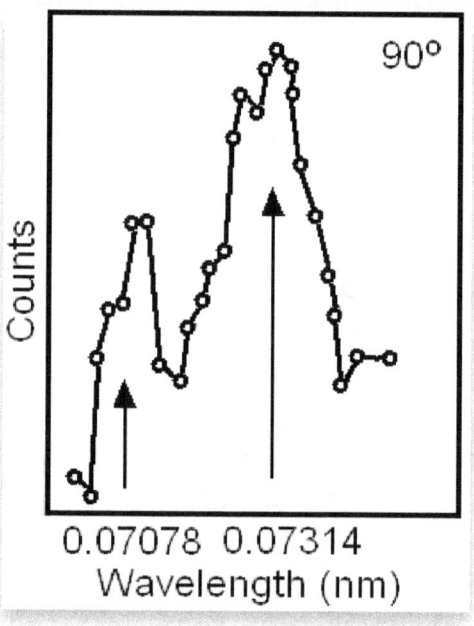

Electrons and photons from Compton and Thomson scatterings can obscure signals from other x ray induced processes as the photoelectric effect and x ray fluorescence (see next section).

Fortunately, the final and initial x ray photons are correlated in energy and can be identified and eliminated if not needed.

Since the factor $(1 - \cos \theta)$ is limited to the range 0 to 2, the λ_{final} is bounded by $\lambda_{\text{initial}} < \lambda_{\text{final}} < \lambda_{\text{initial}} + 2\lambda_C$, the final wavelength will never differ from the initial one for more than 0.04852 Å. The figure shows the relative intensity of the x rays produced by Compton and Thomson scatterings of K_α and K_β lines of a Rh target, as a function of their energies.

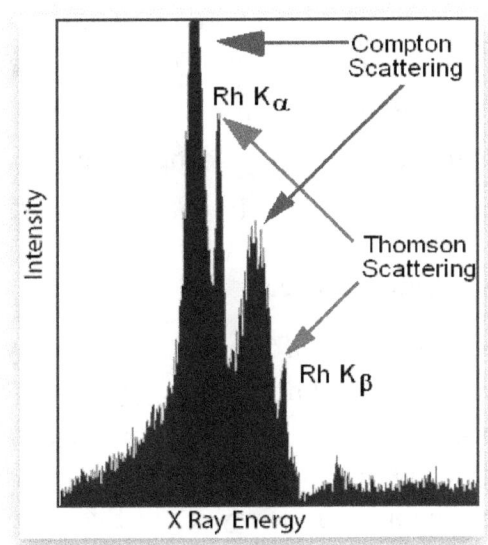

Exercise 2.4

X rays from a Rh target are being Compton scattered by atomic electrons, determine if it is possible that upon scattering the Rh $K_{\alpha 1}$ x ray (energy 20216 eV) and the $K_{\alpha 2}$ x ray (20073 eV) can end up with equal wavelengths. At what angles would this happen?

Solution

The Rh $K_{\alpha 1}$ x ray has a wavelength $E = hf = hc/\lambda \Rightarrow$

$\lambda = hc/E = (1240 \text{ eVnm})/(20073 \text{ eV}) = 6.177 \times 10^{-11} \text{ m} = 0.6177 \text{ Å}$

and through Compton scattering its wavelength shifts to any value in the range 0.6177 Å to 0.6177 + 0.04852 Å = 0.666 Å.

The $K_{\alpha 2}$ x ray has a wavelength of

$\lambda = hc/E = (1240 \text{ eV nm})/(20216 \text{ eV}) = 0.6133 \text{ Å}$,

and by Compton scattering it can reach wavelengths in the range 0.6133 Å to 0.66182 Å; thus the overlap can happen in the range 0.6177 Å to 0.65852 Å.

The value of 0.6177 Å would occur if the $K_{\alpha1}$ x ray "scatters" at $0°$ while the $K_{\alpha2}$ x ray scatters at an angle of

$$\theta = \cos^{-1}[1 - (\lambda_{final} - \lambda_{initial})/(h/m_oc)]$$
$$= \cos^{-1}[1 - (0.6177 - 0.6133)/(0.02426)]=35.05°.$$

The value of 0.66182 Å would occur if the $K_{\alpha2}$ x ray "scatters" at $0°$ while the $K_{\alpha1}$ x ray scatters at an angle of

$$\theta = \cos^{-1}[1 - (0.65852 - 0.6177)/(0.02426)]=144.94°.$$

X ray fluorescence

The production of the K_{α} and K_{β} x rays in Exercise 2.2 are examples of what is known as *x ray fluorescence* (XRF). As depicted in the graphics, a photon kicks an electron out of an element, and a higher energy electron jumps to the lower energy level emitting an x ray photon; in a way x ray fluorescence is a photoelectric effect followed by an spontaneous emission of a photon.

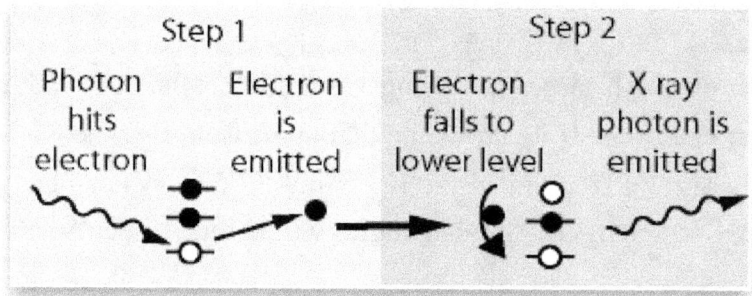

Knowing the energy levels of the different atoms allows us to use XRF as a technique to identify the emitting element by comparing the x ray energy to energy level differences. To identify these peaks different nomenclatures of the transitions have been developed; the figure shows the equivalency between the Siegbahn and IUPAC notations. More about the use of XRF as a technique to study surfaces will be presented in Chapter 3.

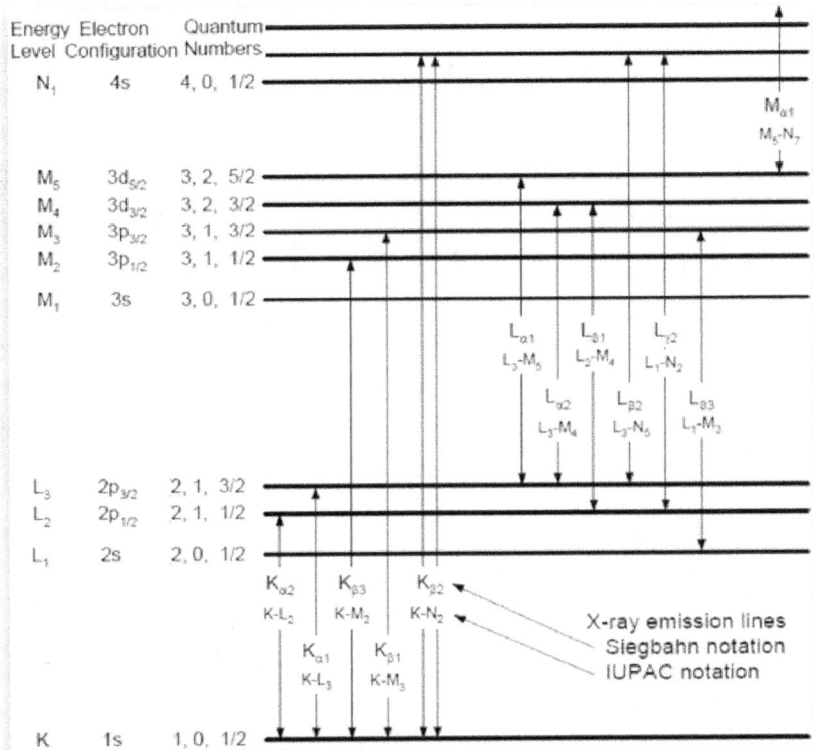

Energy Level: N₁ — Electron Configuration: 4s — Quantum Numbers: 4, 0, 1/2

M₅ 3d₅/₂ 3, 2, 5/2
M₄ 3d₃/₂ 3, 2, 3/2
M₃ 3p₃/₂ 3, 1, 3/2
M₂ 3p₁/₂ 3, 1, 1/2
M₁ 3s 3, 0, 1/2

L₃ 2p₃/₂ 2, 1, 3/2
L₂ 2p₁/₂ 2, 1, 1/2
L₁ 2s 2, 0, 1/2

K 1s 1, 0, 1/2

Exercise 2.5

X rays can be produced by bombarding electrons on a Cd target.

A) Use the Cd energy levels provided in the table to predict the energy of several x ray photons that will be produced.

B) Actual data tend to show an 88 keV photon, could this be produced by the same mechanism as those photons in part A)?

Solution

Electron binding energies for Cd (eV)		
K	1s	26711
Lɪ	2s	4018
Lɪɪ	2p1/2	3727
Lɪɪɪ	2p3/2	3538
Mɪ	3s	772
Mɪɪ	3p1/2	652.6
Mɪɪɪ	3p3/2	618.4
Mɪᵥ	3d3/2	411.9
Mᵥ	3d5/2	405.2
Nɪ	4s	109.8

A)

- $K_{\alpha 1}$: L₁ to K transition: 26.711 - 4.018 = 22.693 keV
- $K_{\alpha 2}$: L₂ to K transition: 26.711 – 3.727 = 22.984 keV
- $K_{\alpha 3}$: L₃ to K transition: 26.711 - 3.538 = 23.173 keV

27

- $K_{\beta1}$: M_1 to K transition: $26.711 - 0.772 = 25.939$ keV
- $K_{\beta2}$: M_2 to K transition: $26.711 - 0.6526 = 26.058$ keV
- Etc.

B) The 88 keV peak cannot be produced by an internal electronic transition in Cd. This is a phenomenon of nuclear origin: ^{109}Cd decays producing gammas of 213.8 keV into ^{109}Ag, which then decays yielding likewise a gamma emission at 88.03 keV

X ray detectors

X rays are detected by means of the ionization they produce on different elements.

Solid-state detectors are made of a disk of Si or Ge doped with Li and with a potential applied across the surfaces. When an x ray photon hits the detector, it produces electron-hole pairs proportional to its energy; these electrons and holes are attracted to the opposite surfaces and constitute an electrical pulse.

Gas filled x ray detectors contain a gas in cylinder with an axial wire at a large voltage (e.g. 1,700 V); photons produce ionizing electrons which in turn produce an avalanche of electrons that is collected by the anode wire as a pulse; the charge collected is proportional to the x ray photon energy.

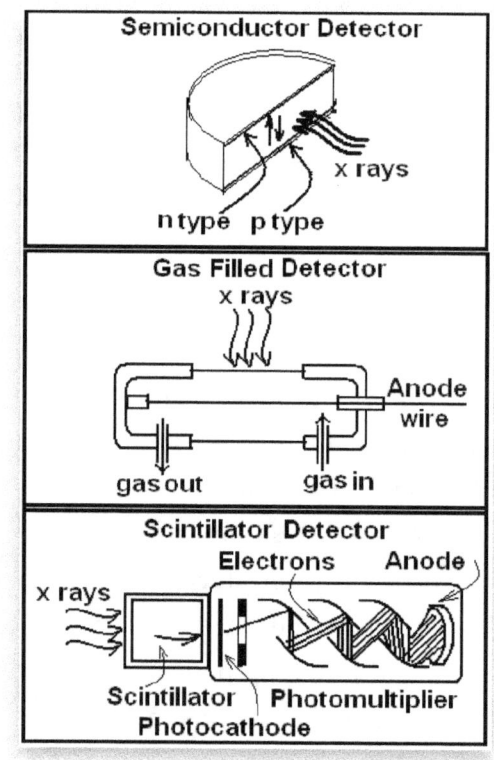

Scintillation counters are crystals of sodium iodide coated with thallium atoms that produce flashes when hit by an x ray photon with the intensity of the flash proportional to the energy of the x ray. The flash in turn goes into a photo multiplier where it releases a number of electrons from a cathode proportional to the flash intensity; the electrons finally produce an avalanche of more electrons inside the photomultiplier.

When the electrons are collected as a current pulse, the signal is quantified and counted according to its magnitude in the *multichannel analyzer*; instrument used to produce the x ray energy spectrum.

Bragg diffraction

Diffraction, in a classical sense, is the bending of waves around edges, at a microscopic level, however, the picture is quite different. When an EM wave hits an atom, the wave's electric and magnetic fields set the atom into vibration at the same frequency, motion which in turn will produce EM radiation which will be irradiated in all directions. If the incident radiation forms a train of waves, it may set many atoms to vibrate simultaneously, producing interference amongst the radiated waves which, at some angles, can be fully constructive or destructive.

This phenomenon can be described using ray diagrams as shown in the picture, if the difference in optical path between the two rays is a multiple of a wavelength, the two rays will be in phase augmenting the light intensity "reflected" (i.e. re-radiated) at that angle. That is, the condition for maximum intensity is

$$n\lambda = 2d \sin \theta,$$

which is known as Bragg's law in honor of English physicists Sir W.H. Bragg and Sir W.L. Bragg (father and son) who, in 1913, explained the "reflection" of x rays by crystals, and, in 1915, shared the Nobel prize. As shown in the diagram, d is the interatomic distance and n is an integer 1, 2, ... This effect gives direct evidence of the atomic structure of crystals and provides the basis for the analysis technique known as x ray diffraction (XRD).

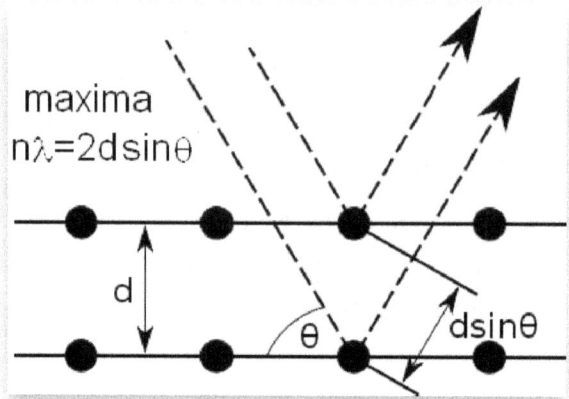

In practical terms, XRD helps to measure the spacing between atomic layers, determine the structure and orientation of a single crystal, as well as identify small crystalline regions. As illustrated in the figure, a usual apparatus consists of an x ray source, sample holder and an x ray detector.

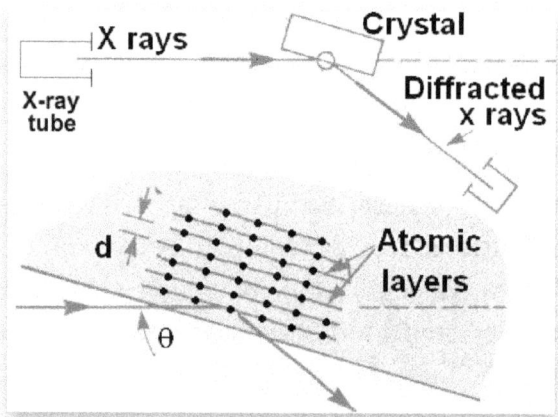

The spectrum produced by an XRD device usually shows the intensity of x rays detected at a given angle; it is customary to plot intensity versus twice the angle due to the set up of the detector-x ray source goniometer. The following figure shows a typical spectrum.

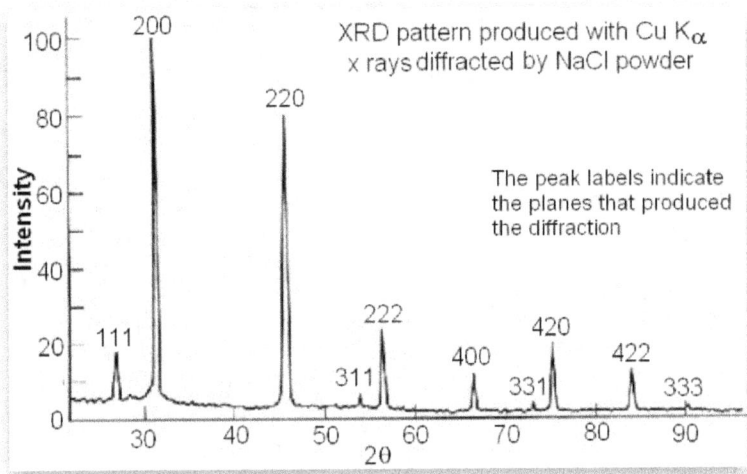

XRD pattern produced with Cu K_α
x rays diffracted by NaCl powder

The peak labels indicate
the planes that produced
the diffraction

Taking advantage of the wave behavior of particles, diffraction is now used with particle beams such as electrons, protons and neutrons; with the advent of spallation neutron sources, there are modern large facilities where slow neutrons are used as a major tool to study materials.

Exercise 2.6

Determine the spacing between layers of NaCl using the approximate value of the ten peaks presented in the previous XRD spectrum.

Solution:

In general, setting n = 1, the Bragg condition is $\lambda = 2d \sin \theta$, using $\lambda = 0.154$ nm we obtain:

111: $d = \lambda/2 \sin \theta = 0.154/[2 \sin (28°/2)] = 0.3182$ nm

200: $d = \lambda/2 \sin \theta = 0.154/[2 \sin (32°/2)] = 0.279$ nm

220: $d = \lambda/2 \sin \theta = 0.154/[2 \sin (46°/2)] = 0.197$ nm

311: $d = \lambda/2 \sin \theta = 0.154/[2 \sin (54°/2)] = 0.169$ nm

222: $d = \lambda/2 \sin \theta = 0.154/[2 \sin (56°/2)] = 0.164$ nm

400: $d = \lambda/2 \sin \theta = 0.154/[2 \sin (66°/2)] = 0.141$ nm

331: $d = \lambda/2 \sin \theta = 0.154/[2 \sin (74°/2)] = 0.1279$ nm

420: $d = \lambda/2 \sin \theta = 0.154/[2 \sin (76°/2)] = 0.125$ nm

422: d = λ/2 sin θ = 0.154/[2 sin (84°/2)] = 0.115 nm

333: d = λ/2 sin θ = 0.154/[2 sin (90°/2)] = 0.1088 nm

Exercise 2.7

An effect similar to diffraction takes place when atoms go through a crystal or periodic nano structure. One example of such effect is presented in the figure (adapted from Schattenburg

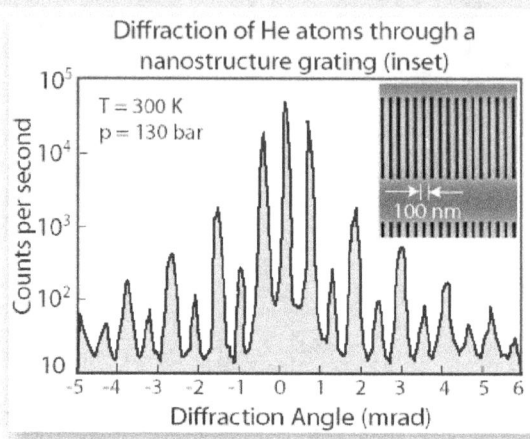

et al., and Savas et al.) which was produced by He atoms at 300 K passing through the slots (dark regions) of a nanostructure grating made of silicon nitride bars spaced 100 nm; the oscillatory behavior of the intensity is characteristic of wave phenomena. Determine

A) the wavelength of the He atoms,

B) the location of the first maxima which for transmission is given by sin θ = m λ/d, and

C) How does the wavelength of the He atoms compares to the wavelength of the laser light (e.g. 633 nm) and to the interatomic distance of, say body-centered cubic iron (0.287 nm)?

Solution:

A) λ = h/p, p = mv, but since E = mv²/2 = p²/2m ⇒ p = [2mE]$^{1/2}$, but since E = 3kT/2 then

p = [3mkT]$^{1/2}$ = [3×(4.002602u×1.66053886×10^{-27} kg/u)×

(1.3806503×10^{-23} m² kg s^{-2}K^{-1})×300 K]$^{1/2}$ = 9.08×10^{-24} kg m/s

Then λ = h/p = (6.626068×10^{-34} m² kg/s)/(9.08×10^{-24} kg m/s)

= 7.2911×10^{-11} m = 0.072911 nm.

B) The angle for a m = 1 maximum is

$\sin\theta = m\lambda/d = 1 \times 7.2911 \times 10^{-11} m/100 nm = 7.2911 \times 10^{-4}$

$$\Rightarrow \theta_1 = 0.0417° = 0.727 \text{ mrad}$$

in close agreement with the first peak to the right of the center.

C) Comparing the He wavelength to that of laser light of 633 nm, with a wavelength of 0.072911 nm, the He beam has a wavelength about 8,600 times "smaller" than the laser light, but it is only about 4 times "smaller" than the interatomic distance of BCC iron of 0.287 nm.

Further reading

Like the concepts in the previous chapter, those in the present chapter are explained in detail in modern physics books. Again, we direct readers to the textbooks of Beiser, Krane, Serway and Tipler listed in the Bibliography. For an online comprehensive site for physics concepts, we recommend Hypherphysics, for an online activity on Bragg diffraction we recommend Glenn A. Richard's web site, and for wealth of information including physical constants and atomic weights of elements we direct readers to the web site of the National Institute of Standards and Technology, NIST. Detailed information about x ray production by synchrotron can be found at, e.g. Stanford Linear Accelerator; again full listings can be found in the annotated bibliography for Chapter 2 at the end of the book.

Problems

Problem 2.1

If we have that the first ionization of Helium is 2372.3 kJ/mol and its second ionization energy is 5250.5 kJ/mol, what would be the temperature of a sea of free electrons if they were to ionize "ionized Helium"? Recall that $1kJ/mol = 1.04 \times 10^{-2}$ eV/particle.

Problem 2.2

Electronic transitions between levels occur only between levels in which $\Delta l = \pm 1$. Then, recalling that $l = 0$ (s), $l = 1$ (p), $l = 2$ (d), etc., and that $K = 1$, $L=2$, draw the allowed transitions between the energy levels of Al shown in the diagram,

72.55 eV	$L_3\, 2p_{3/2}$
72.95 eV	$L_2\, 2p_{1/2}$
117.8 eV	$L_1\, 2s$
1559.6 eV	$K_1\, 1s$

and estimate the energy of the photons emitted during those transitions.

Problem 2.3

In a photoelectric experiment a metal is illuminated with light of wavelength 470 nm and the stopping potential (i.e. the potential needed to stop the most energetic electrons) is found to be 0.34 V.

A) What is the metal being analyzed?

B) If irradiated with light of wavelength 3.3×10^2 nm, what stopping potential will be needed?

Problem 2.4

X rays from an Ag target are being Compton scattered by electrons.
A) Determine if it is possible that upon scattering the Ag $K_{\alpha 1}$ x ray
(energy 22162.92 eV) and the $K_{\alpha 2}$ x ray (21990.3 eV) can end up
with equal wavelengths.
B) At what angles would this happen?

Problem 2.5

The figure shows the spectrum of an x ray pyroelectric generator
that uses a crystal of $LiTaO_3$ and a Cu target. There are two cycles
of operation, a warm up and a cool down. In one cycle electrons hit
the Ta in the crystal yielding characteristic radiation, and in the
next cycle electrons hit the Cu target producing characteristic
radiation. Identify the radiation produced in the spectrum shown
using the table shown.

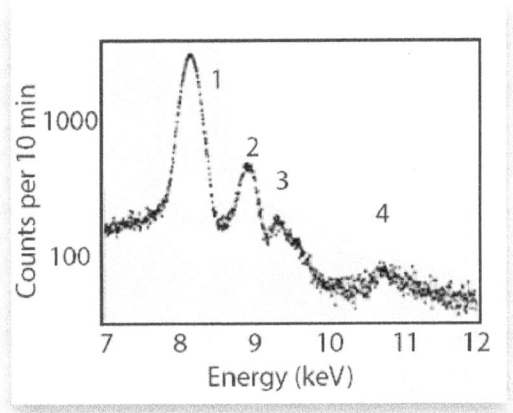

	Kα1	Kα2	Kβ1	Lα1	Lα2	Lβ1	Lβ2	Lγ1	Mα1
Cu	8047	8027	8905	929	929	949.8			
Ta	57532	56277	65223	8146	8087	9343	9651	10895	1710

Problem 2.6

With the diffraction pattern shown in the figure, estimate the d-spacing in GeO_2. The structure has a tetragonal structure, with a unit cell size of a = b = 0.44 nm, c = 0.29 nm. The wavelength used is 0.154 nm, the figure displays the 2θ angle and the Miller indices of each peak.

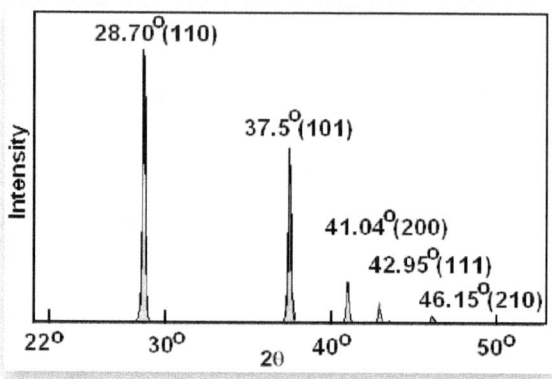

Problem 2.7

Taking the same situation as in Exercise 2.7, consider that the temperature of the He atoms is reduced to 250 K, but the nanostructure keeps the same pitch of 100 nm.

Again, determine the following:

A) The wavelength of the He atoms.

B) The location of the first order maxima.

C) How do the results in (A) and (B) above, compare to those of Exercise 2.7?

Chapter 3: X Ray Fluorescence

Glocker and Schreiber first proposed the use of XRF as an element identification technique in 1928. In this chapter we will review the implementation of this technique for surface analysis. XRF is used for non-destructive elemental analyses of solids such as rocks, minerals, sediments, pipes, and even works of art as shown in the picture of UCSD's materials scientist Maribel López performing an XRF study of Murillo's painting "The Nativity" at Houston's Museum of Fine Art.

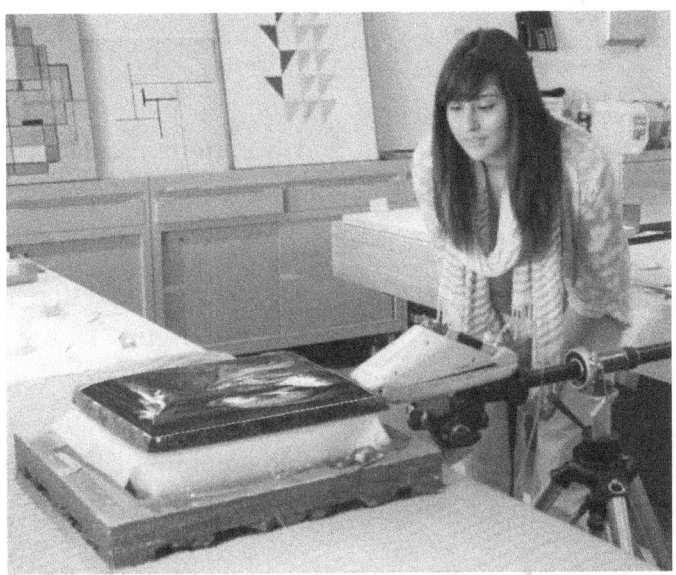

As explained before, the identification of elements in solids is possible by the ionization of an inner electron of the atom under study by means of x rays, followed by the decay of an outer electron replacing the missing inner electron accompanied with the release of what is termed fluorescent x ray radiation; the word "fluorescence" refers to emission of EM radiation by a substance, at a wavelength (or wavelengths), different from the incident wavelength of EM radiation.

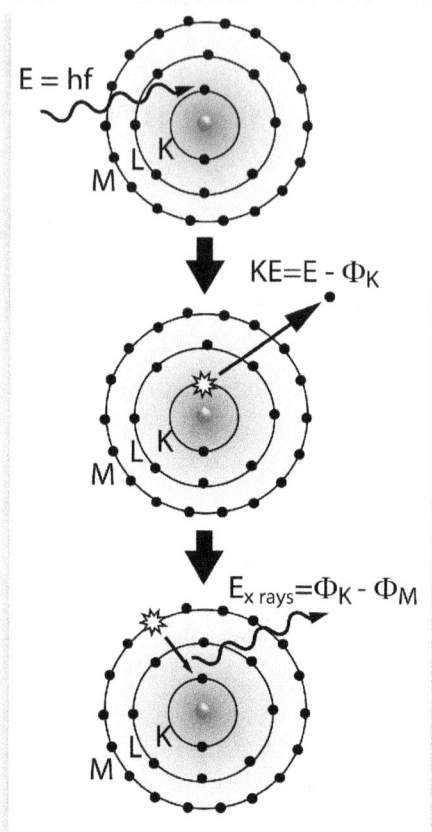

Since the energy of the emitted photon is equal to the energy difference between the involved orbitals, the resulting fluorescent x rays can be used to identify the emitting element. Similarly, the intensity of the x rays can be used to gauge the relative abundance of elements present in the sample, although the exact value of concentrations (e.g. parts per million, etc.) must be obtained by comparison to calibration standards whose composition is known from other techniques.

As an example, the accompanying figure shows the x ray fluorescence intensity as a function of the x ray energies obtained by irradiating the Pueblo Indian piece of ceramic shown. The table A2.1 in Appendix 2 lists the most common XRF lines for a variety of elements.

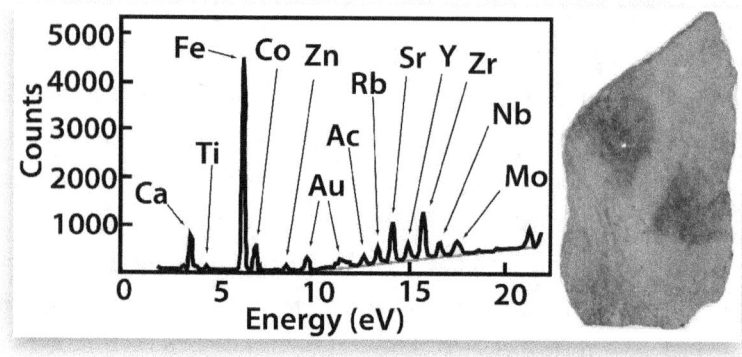

XRF Spectrometers

In practice, a sample is illuminated by an intense x ray beam (incident beam), typically produced from Ag, Rh, W, Cu, Mo, or Cr target. Conventional x ray generators in the range of about 20 to 60 kV are used to produce a continuous bremsstrahlung radiation as well as the characteristic K and L emission lines of the target.

Spectrometers come in all shapes and sizes; the accompanying figure shows a diagram of a portable handheld XRF gun. Current models operate with x rays produced with an Rh or Ag target at an operating voltage of 40 kV and use a solid state Si PIN detector.

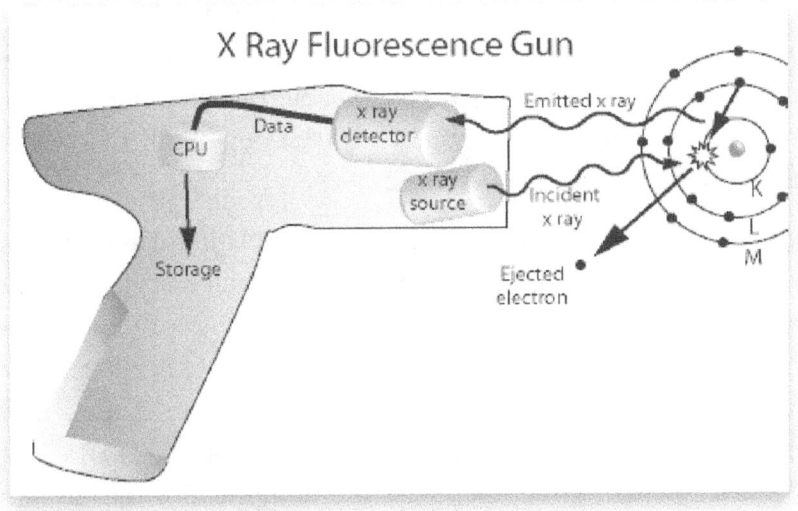

Characteristics of XRF

Since the fluorescent x rays emitted are produced by the elements present in the sample, and since each element can produce photons of different energies (e.g. K_α, K_β, etc.) a proper identification of the x rays is necessary.

The detection of XRF photons is difficult for many reasons. First and foremost, the fluorescence process is inefficient, i.e. it produces relatively few photons, and second, if the produced radiation is of low energy it will be attenuated by air on its way to the detector. Some spectrometers keep the sample-to-detector path under vacuum (about 10 Pa, a level reachable by a roughing pump of the rotary type, see Appendix 1). For less demanding measurements a vacuum of about 100 Pa, can be easily achieved with a diaphragm pump; as in the case of the fluorescence gun shown above.

X-Ray fluorescence is particularly well-suited for investigations that involve bulk analyses of major elements (Si, Ti, Al, Fe, Mn, Mg, Ca, Na, K, P) as well as analyses of trace elements (>1 ppm; Ba, Ce, Co, Cr, Cu, Ga, La, Nb, Ni, Rb, Sc, Sr, Rh, U, V, Y, Zr, Zn). In principle, XRF could analyze elements as light as beryllium (Z=4), but their low yields make it difficult to study elements lighter than sodium (Z=11). XRF is capable of detecting elements in concentrations from PPM range to 100%; depths of the order of μm can be analyzed (see Appendix 2).

A major drawback of XRF is the inability of distinguishing among isotopes of an element or elements in different valence states. Other complications include the possibility of emission of Auger electrons (see Chapter 4) during XRF excitation. Other limitations are the range of energy and wavelengths XRF can detect, usual ranges are from 0.11 keV to 60 keV for energy, and from 11.3 nm to 0.02 nm for λ. Yet another possible complication is the slow velocity of data collection, with intensity values of thousands of counts per second, and exposition times which usually run in the minutes and up to an hour. An additional complication may be the availability of standard samples, particularly for quantification purposes.

Sample preparation

Perhaps the best advantage of XRF over wet chemical methods, which also disclose elemental composition, is the minimum need for sample preparation. Most solid materials do not require any preparation as long as the sample is of appropriate size (relative to the x ray beam) and has a clean and flat surface. Typically samples of at least several grams of material are needed. The beam of x rays used generally covers areas of lengths of microns to millimeters.

Heterogeneous material, such as polycrystalline rocks and pottery may require to be irradiated in several areas for an accurate average composition; painted surfaces will also require special attention. For powders, such as soils, etc., it is common to prepare samples by pelletizing, i.e. by grinding and homogenizing the material, and pressing into a solid with a smooth surface and good solid stability. Particle size is important and, in general, must be smaller than 50 μm. Surface elements can present additional complications due to oxidation. Surface roughness of the sample can distort the signal and yield misleading results.

Depth of XRF

In general, commercial XRF apparatus will have a penetration of up to millimeters in solids; for deeper x ray penetration higher photon energies are needed. However, for homogeneous samples penetration is not a major issue as the layers close to the surface will have similar compositions than deeper layers; it certainly would be relevant to assess homogeneity by an alternative technique.

The figure A3.4 in Appendix 3 (adapted from Evans Analytical Group) shows the typical analysis depth of XRF compared to other techniques.

The detection of buried elements presents a more serious problem, both for the need of more energetic x rays to produce fluorescence as well as for the possible absorption by the medium of the emitted fluorescence. Absorption is of special importance for lighter elements, which tend to produce lower energy fluorescence. The table A2.3 in Appendix 2 lists the layer thickness from which 90% of the fluorescence is produced in graphite, glass, iron or lead for a variety of x ray sources. For instance, when irradiating a sample of glass, containing tungsten impurities, the glass (actually the tungsten impurities) will produce fluorescence in a layer of thickness 429 μm. By comparison, a lead sample, containing tungsten impurities, will only yield information about the impurities from the top 22.4 μm.

XAFS

A technique related to XRF is the one known as *X Ray Absorption Fine Structure* (XAFS). Since the probability of absorption of x rays by core electrons is very sensitive to the chemical and physical state of the atom, XAFS can provide information about the oxidation, local atomic structure of the material (coordination number and molecular orientation). To achieve this, XAFS requires higher energy resolution than XRF, while standard energy dispersive detectors can measure emission lines with a resolution of, say, 150 eV, XAFS requires resolutions of the order of 1 eV, comparable in magnitude to chemical shifts. XAFS explores the energy range between the core level and unoccupied orbitals, range known as the *absorption edge*, typically extending from about 100 eV below the edge up to about 1000 eV above it.

Either a fluorescent or transmission configuration can serve to collect data using XAFS and such variations give rise to other more specialized tools like x ray absorption near edge structure (XANES), extended absorption near edge structure (EXAFS), near-edge energy (NEXAFS) and surface extended x ray absorption fine structure (SEXAFS). Properly speaking these techniques provide bulk information but also operate on the surface. XAFS and its offspring techniques will not be discussed any further here, interested readers are directed to the excellent books and articles of Newville, Konigsberger and Prins, Rehr and Albers and Stoehr, the sites Surface Analysis Forum and XAFS.ORG are also valuable sources.

Exercise 3.1

A silicon wafer was treated with a process known as siliciding producing a layer of $WFe_6 + SiCl_2H_2$. If XRF with hard x ray photons of 15 keV is used to examine the layer, what would be the energy of some of the x rays emitted by tungsten? Label the x rays according to the proper Siegbahn nomenclature (e.g. $K_{\alpha1}$, etc.).

Level	Energy (eV)
K 1s	69525
L1 2s	12100
L2 2p1/2	11544
L3 2p3/2	10207
M1 3s	2820
M2 3p1/2	2575
M3 3p3/2	2281
M4 3d3/2	1949
M5 3d5/2	1809
N14s	594.1
N2 4p1/2	490.4
N3 4p3/2	423.6
N4 4d3/2	255.9
N5 4d5/2	243.5
N6 4f5/2	33.6
N7 4f7/2	31.4
O1 5s	75.6
O2 5p1/2	45.3
O3 5p3/2	36.8

Energy levels of W (keV)

M4 1.95

M3 2.28

M2 2.58

M1 2.82

Lower levels

L3 10.2

L2 11.5

L1 12.1

K 69.5

Solution

The incident x ray is not energetic enough as to produce K x rays, but it can produce L x rays:

- $L_{\alpha1}$: L_3–M_5 $\approx$ 10.207–1.809 = 8.398 keV
- $L_{\alpha2}$: L_3–M_4 = 10.207–1.949 = 8.258 keV
- $L_{\beta1}$: L_2–M_4 = 11.544–1.949 = 9.595 keV, Etc.

Exercise 3.2: Calibration of XRF

As an example on how to calibrate XRF data, G.D. Rusche studied a sample of zonolite found in attic insulation. XRF was used to determine the Ba, Cr and V concentration using three calibrated standards from NIST, BCR and JG. The results were then compared to a sample from a near mine to determine if the attic sample came from such source. The concentrations of the calibrated samples are shown in the first table, and the XRF counts obtained from the calibrated samples and the attic and mine samples are presented in the second table.

A) Determine the actual concentrations of Ba, Cr and V in the attic and mine samples.

B) Is the attic sample compatible with the mine sample?

Element	Source	Concentration (µg/l)
Ba	NIST	462
	BCR	681
	JG	1254
Cr	NIST	1990
	BCR	962
	JG	281
V	NIST	122
	BCR	194
	JG	39

Element	Source	XRF counts
Ba	NIST	66
	BCR	90
	JG	179
	Mine	158
	Attic	185
Cr	NIST	2486
	BCR	1150
	JG	356
	Mine	2046
	Attic	356
V	NIST	510
	BCR	890
	JG	156
	Mine	780
	Attic	850

Solution

A) Using the data from the calibrated three samples, one can plot the x ray count vs. concentration for each element to obtain the lines shown in the figure. Using straight line fits for each element, one can use them to predict the concentration of the attic and mine samples, the results are shown

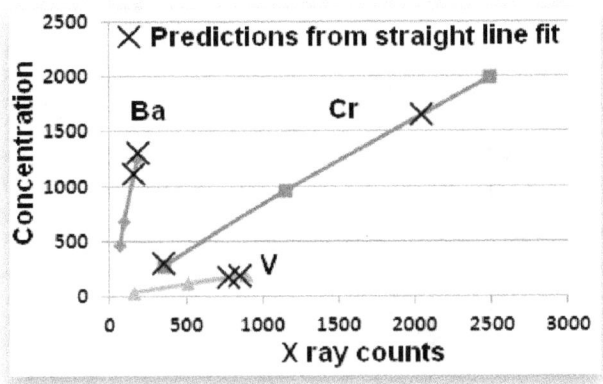

in the plot as "×" and their values (in μg/l) are:

    Attic Mine

Ba: 1117 1302

Cr: 1649 299

V: 173 188

B) To determine compatibility between the attic and mine samples one can look at the relative proportions of concentrations of elements:

	Attic	Mine
Ba/Cr	4.35	0.67
Ba/V	6.91	6.43
Cr/V	1.59	9.5

Undoubtedly, the attic sample is not compatible with the mine sample and it probably has a different origin.

Exercise 3.3 Relative concentrations

XRF is widely used in speciation analysis of soil samples; unfortunately, as soils vary in their composition is hard to obtain exact concentrations as there are no calibrated samples. One way of obtaining relative concentrations is by direct comparison of the areas under the intensity peaks.

The figure (adapted from Baranowski et al.) shows the XRF peaks of Al and Si of soil obtained from Silesia in Poland. The vertical

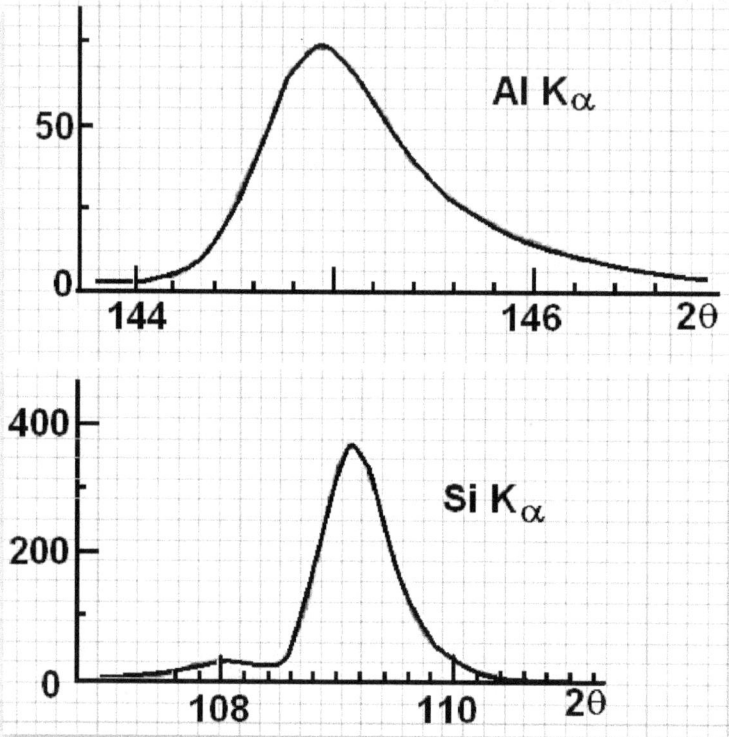

axis indicates counts and the collection time is the same in both profiles.

A) Use this information to estimate the relative concentration of Al and Si. [Notice that the researchers used a crystal to diffract the x rays according to their wavelength (inversely to their energy) and thus the curves plot intensity (counts) versus twice the angle 2θ (degrees).]

B) Using other methods, calibrated samples of soils were obtained and yielded concentrations of 65.12 mg of Al and 275.36 mg of Si per gram of soil. Does your answer to part A) matches these experimental results?

Solution.

Drawing the background baseline under the curve, the area under the Al peak encloses approximately 90 squares, since the area of each Al square is $\approx (2/19) \times (50/8)$, the total area in counts×degrees is 59.21.

Likewise, the area under the Si peak encloses approximately 47

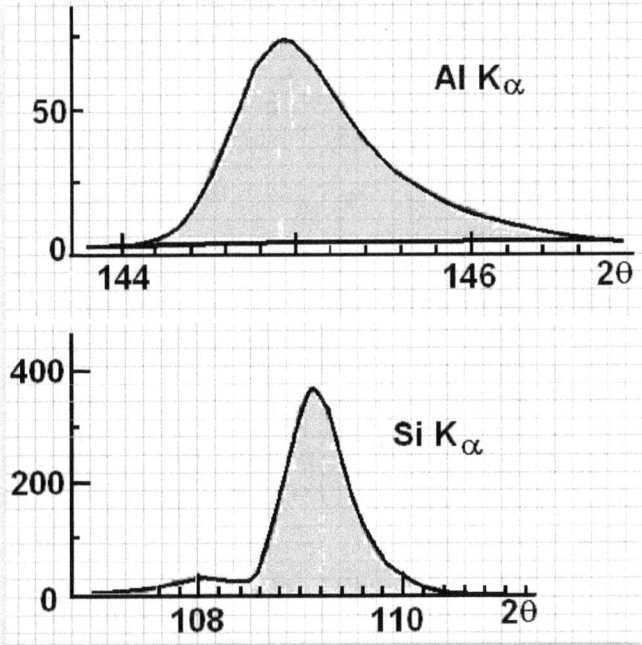

squares, since the area of each square is $\approx (2/11.5) \times (400/13)$, the total area is 251.50 counts×degrees.

Thus the relative concentration of Si:Al is $251.5/59.21 = 4.24$, which compares well to the expected experimental ratio of $275.36/65.12 = 4.22$.

Further reading

Readers interested in more advanced textbooks are directed to the list provided in the section of general sources in the Bibliography. Likewise, the list of web sites provided specifically for XRF include tables of x ray emission energies, fluorescent yields, an applet and even a link to a request form for a free x ray data booklet from the Lawrence Berkeley National Laboratory.

Problems

Problem 3.1

The spectrum shown in the figure was obtained by Linke et al. by exposing a coin to Rh radiation, with an acceleration voltage of 50kV and a filter of palladium. Based on the elements identified on the chart, label the x ray lines observed using the Siegbahn nomenclature. Assume that the detector used had a resolution of 155 eV at the Mn K-alpha line.

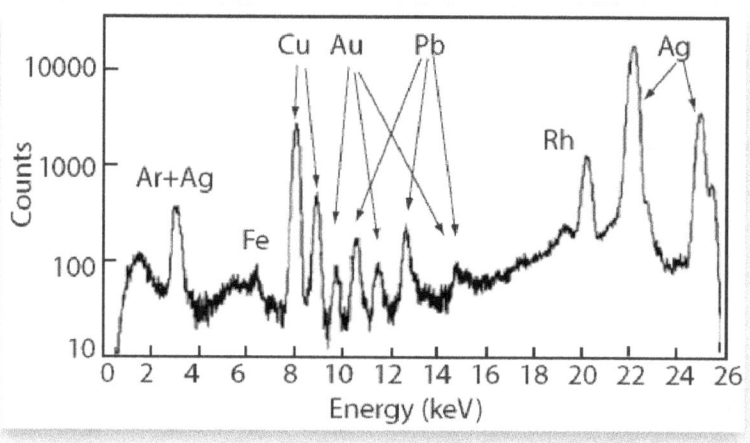

Problem 3.2

Given the same information as in Exercise 3.2, i.e. the information about element, source, and calibration for Ba, Cr, and V, as well as their XRF counts (NIST, BCR, and JG standards), discuss the results for two

	XRF Counts	
	Sample 1	**Sample 2**
Ba	221	73
Cr	2700	1805
V	1210	690

other samples (Sample 1 and Sample 2), that also contain Ba, Cr, and V with the count rate shown in the table.

Problem 3.3

Repeat the study of Exercise 3.3 and use the graph of intensity versus 2θ to estimate the relative concentrations of K and Ca found by XRF in the Silesia soil. For these elements, the concentrations obtained from the calibrated samples were 16.53 mg of Ca and 8.81 mg of K per gram of soil.

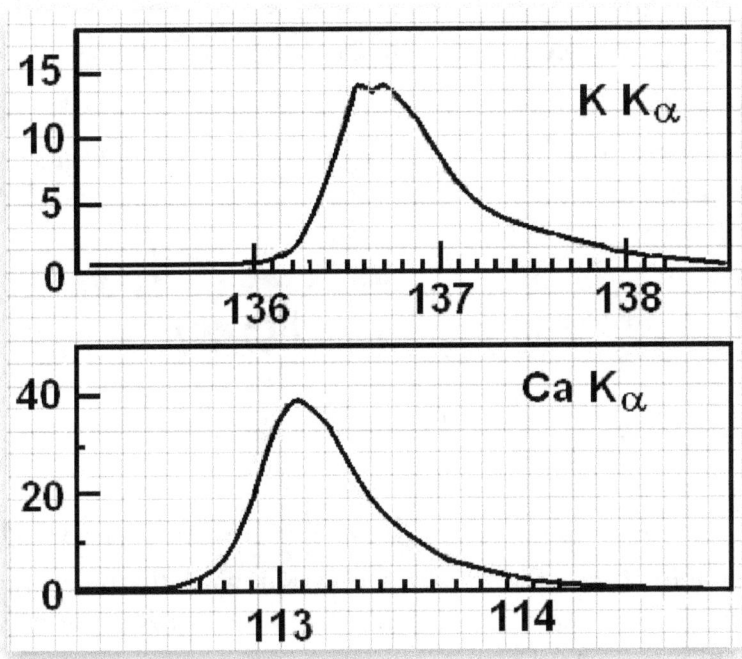

Chapter 4: X Ray Photoelectron Spectroscopy

X ray photoelectron spectroscopy (XPS) is the implementation of the photoelectric effect to the study of materials. It was developed in the 1960s by the Swedish Kai Siegbahn who earned the Nobel Prize in 1981 for his work; the technique is also known historically as Electron Spectroscopy for Chemical Analysis (ESCA).

As explained in the section "Photoelectric effect" and depicted in the figure, in XPS an incident x ray (of energy hf, f being the frequency) knocks an electron (with binding energy BE) out of the atom which escapes with an energy

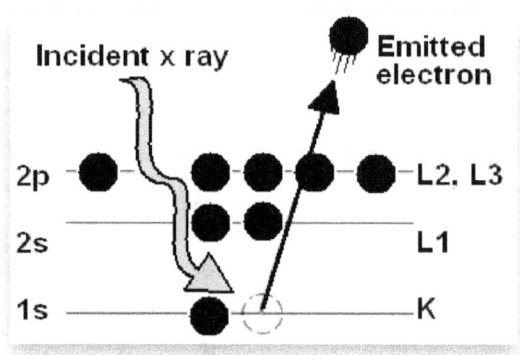

equal to $E = hf - BE - \Phi$, where Φ is the work function of the spectrometer (of the order of a few eV). Knowing Φ, the energy of the incident x ray and capturing the electron to measure its kinetic energy allows the determination of the binding energy of the electron and thus the identification of the element. The figure below (from Dr. B. Vincent Crist) shows a beautiful description of the complete XPS process from x ray irradiation to compilation of the electron energy spectra.

XPS is used to study the elemental composition, chemical state and electronic state of a material. Normally it explores the top 10 nm or top 20 layers of a surface. To avoid scattering of the XPS electrons with air, XPS is performed in a ultra high vacuum (UHV) chamber,

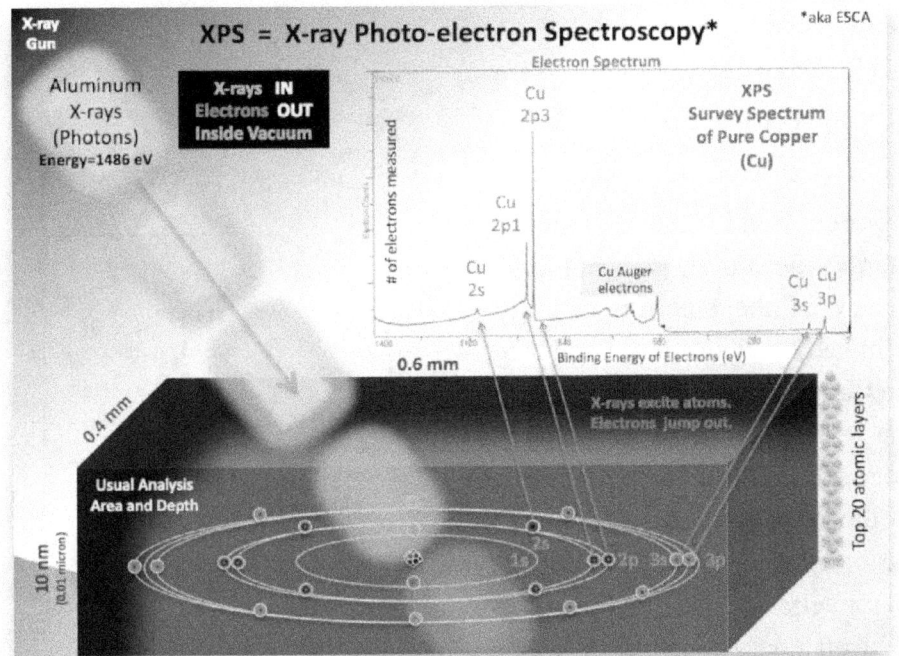

i.e. with pressure less than 10^{-7} Pa. Samples can be studied without any preparation other than normal cleaning, but some applications can benefit from being sputtered with ions to clean off surface contamination. Care should be taken with samples that can become charged since this effect will shift the peak positions; more about this aspect will be discussed later. Also, samples that degas can pose a problem since their composition could change during the experiment.

XPS can detect elements starting from Li (Z=3) and higher; hydrogen (Z = 1) and helium (Z = 2) cannot be detected due to the low probability of electron emission. Detection limits for most of the elements are in the parts per thousand range, but it can be increased to parts per million (ppm) for large concentrations or through long collection times (overnight). XPS is applicable to inorganic compounds, metals, semiconductors, organic material, bio-materials, as well as oils and gases under special conditions. XPS is in general non-destructive and can be safely used in the study of works of art, particularly in recent times, since the probe beam can be focused down to areas of a few squared microns; assuming vacuum compatibility.

XPS Spectra

Once the XPS electrons are captured and their kinetic energy measured, the binding energy of the electron can be estimated and used to produce the XPS spectrum, which is a histogram of the number of electrons captured as a function of their binding energy. Since the binding energy is different for every atomic level, the

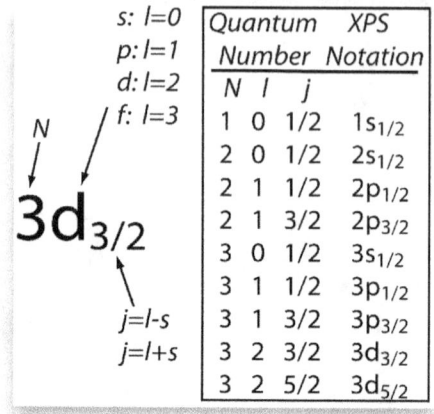

Quantum			XPS
Number			Notation
N	l	j	
1	0	1/2	$1s_{1/2}$
2	0	1/2	$2s_{1/2}$
2	1	1/2	$2p_{1/2}$
2	1	3/2	$2p_{3/2}$
3	0	1/2	$3s_{1/2}$
3	1	1/2	$3p_{1/2}$
3	1	3/2	$3p_{3/2}$
3	2	3/2	$3d_{3/2}$
3	2	5/2	$3d_{5/2}$

electron energy spectrum will show peaks at these energies and, depending on the energy resolution of the electron detector (usually of the order of an eV down to about 1/3 eV), at the energy of the sublevels within shells.

The notation used in XPS to identify the energy levels follows the standards set by quantum mechanics. As mentioned in Chapter 1, the binding energy of an atomic electron depends not only on the energy shell of the level they occupy, but also on the magnetic interaction between its intrinsic spin and the orbital angular momentum. Thus the energy levels are characterized by the quantum numbers of the orbital, and of the total and orbital angular momentum, N, j and l, respectively; these are used in the nomenclature shown in the figure. The accompanying table shows the most common energy states encountered in XPS.

The figure shows an example of XPS spectra (adapted from K. Janssens work) obtained by irradiating an Ag target with Al Kα x rays of 1486.6 eV. As it will be explained in the next section, the captured electrons have different origins; here it suffices to mention

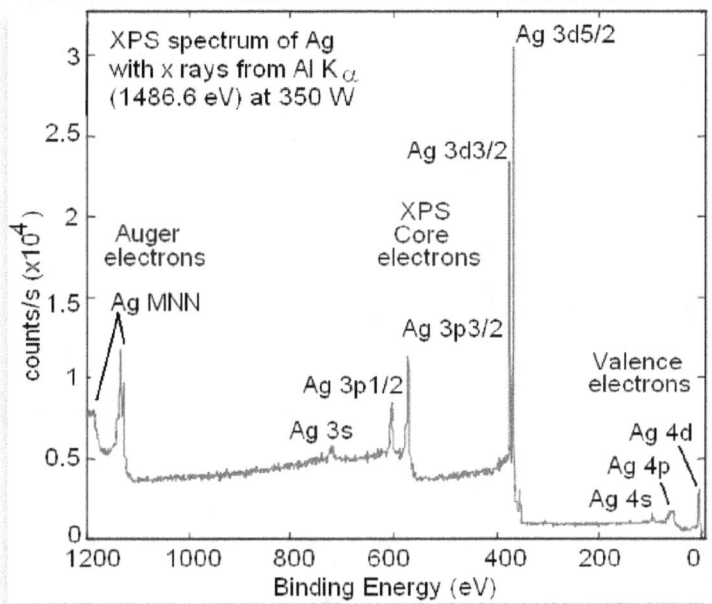

that the main peaks are from the XPS electrons emitted from the core levels.

The background is produced by electrons that scatter inelastically with other atoms on their way out and suffer energy loss; since those electrons arrive at the detector with smaller kinetic energies, they are counted as electrons with higher binding energies thus increasing the noise at higher binding energy and producing the stepped background. The intensity of all peaks is proportional to the intensity of the x ray beam.

The following chart shows the binding energies of the electrons in the different energy levels as a function of the charge (Z) of the atoms. Notice that in all cases the "s" levels have only one curve while there are two "p", "d" or "f" curves; this split of energy levels is due to the interaction energy between the particle spin and their orbital angular momentum.

The energy of a level varies according to the total angular momentum, $j = l + s$. States with $l > 0$ can have two possible j values: $j = 1 - \frac{1}{2}$, $1 + \frac{1}{2}$ for $l = 1$ (p), or $j = 2 - \frac{1}{2}$, or $j = 2 + \frac{1}{2}$ for $l = 2$ (d), etc.

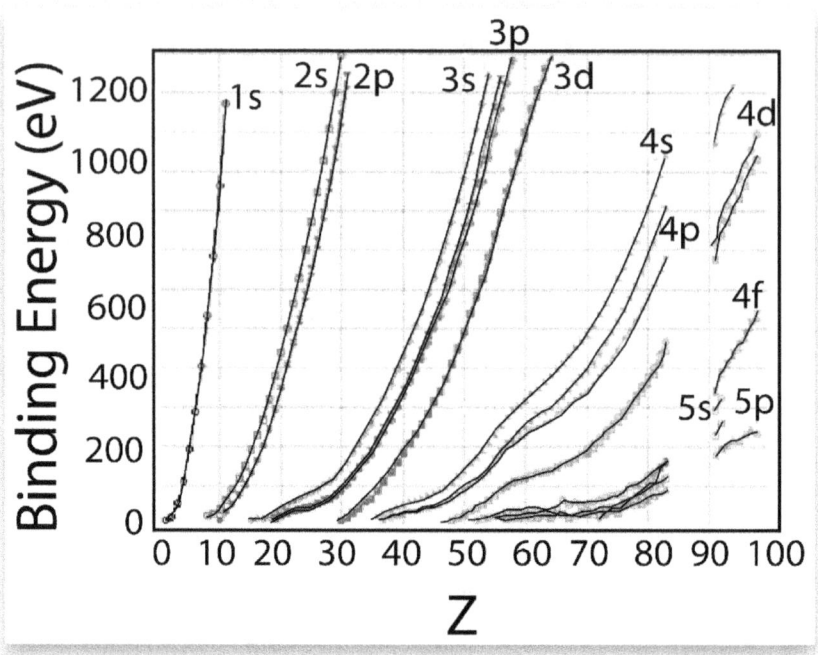

In the XPS spectra this results in pairs of peaks with relatively close values, as illustrated in the following figure (adapted from data of E. Pegg), which shows a superposition of two spectra from Ti and TiO_2. A useful feature of these pairs of peaks is that they have a well defined ratio of intensities. Since different energy levels can be occupied by a different number of electrons, an x ray

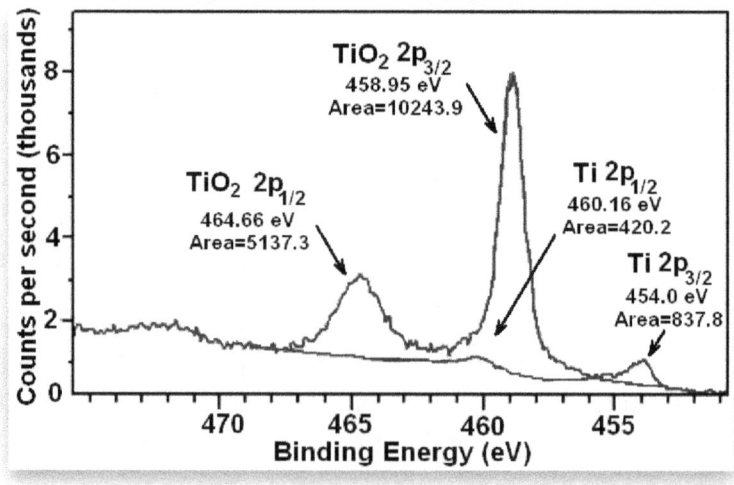

will have a larger probability of hitting an electron in a high occupancy level; thus levels with higher electron occupancy produce more intense XPS peaks. Hence, the ratio of the intensities of a given pair

Orbital l		j	degeneracy	Electron level
1s	0	1/2	1	1s
2s	0	1/2	1	2s
2p	1	1/2	2	$2p_{1/2}$
2p	1	3/2	4	$2p_{3/2}$
3d	2	3/2	4	$3d_{3/2}$
3d	2	5/2	6	$3d_{5/2}$
4f	3	5/2	6	$4f_{5/2}$
4f	3	7/2	8	$4f_{7/2}$

of peaks equals the ratio of the occupancies of such energy levels; such occupancy is given by the *degeneracy* which equals *2j+1*. The accompanying table shows the degeneracy of several energy levels.

For instance, all pairs composed by the 2p3/2 and 2p1/2 peaks will have a degeneracy ratio of 4 to 2 i.e. a ratio of 4/2 = 2. Quantifying the intensity of such peaks in the previous TiO_2 spectrum by the area under the peaks (as measured with respect to the background), we find the ratio of the areas to yield 10243.9/5137.3=1.994. Repeating the estimation with the 2p3/2 and 2p1/2 peaks of Ti we also find a ratio of 837.8/420.2=1.9938 in close agreement with the expected value. In general, all p peaks ($p_{1/2}$, $p_{3/2}$) will have an area ratio of 1:2, d peaks ($d_{3/2}$, $d_{5/2}$) of 2:3, f peaks ($f_{5/2}$, $f_{7/2}$) of 3:4, etc.

Exercise 4.1

In Janssens spectrum an Ag target is irradiated with Al Kα x rays of 1486.6 eV.

A) Determine the kinetic energies the photoelectrons had when they were captured.

Energy levels of Ag (keV)

Level		Energy (eV)
K	1s	25514
L1	2s	3806
L2	2p1/2	3524
L3	2p3/2	3351
M1	3s	719
M2	3p1/2	603.8
M3	3p3/2	573
M4	3d3/2	374
M5	3d5/2	368.3
N1	4s	97
N2	4p1/2	63.7
N3	4p3/2	58.3

M4 0.37
M3 0.57
M2 0.60
M1 0.72

Lower levels

L3 3.35
L2 3.52
L1 3.80

K 25.5

Assume that the spectrometer work function is negligible and use

55

the exact Ag binding energies listed in the table (from www.webelements.com, note that there are two scales in the table, [eV] and [keV], showing basically the same information) or from the Table A2.2 in the Appendix 2.

B) Show that the peaks labeled as Auger peaks cannot indeed be XPS peaks.

Solution

A) Some examples:

Ag $3d_{5/2}$: $E = hf - BE - \Phi = 1486.6 - 368.3 - 0 = 1118.3$ eV

Ag $3p_{3/2}$: $E = hf - BE - \Phi = 1486.6 - 573 - 0 = 913.6$ eV

Ag 4s: $E = hf - BE - \Phi = 1486.6 - 97 - 0 = 1389.6$ eV

Etc. Notice that, by convention, in graphic representations of the spectrum BE increases to the left and the kinetic energy E increases to the right.

B) The two Auger peaks shown appear to have binding energies of, say, 1150 eV and 1180 eV, looking at the table of Ag energies, no levels with those energies are found, consequently the peaks do not correspond to XPS electrons.

Relationship to Auger electrons

As studied before, after the XPS electron is emitted, another electron will occupy its place in the atom emitting a photon in the transition. The latter will escape the material or will in turn kick a second electron out in a process known as Auger effect.

For every XPS electron emitted there will be either a photon or a second (Auger) electron being emitted. Although the emitted photon and Auger electron carry information that can aid in the identification of the element, XPS, strictly speaking, refers only to the analysis based on the first photoelectron emitted. Although the x rays and Auger electrons tend to contaminate the XPS spectra, such signals can be identified and filtered out of the analysis.

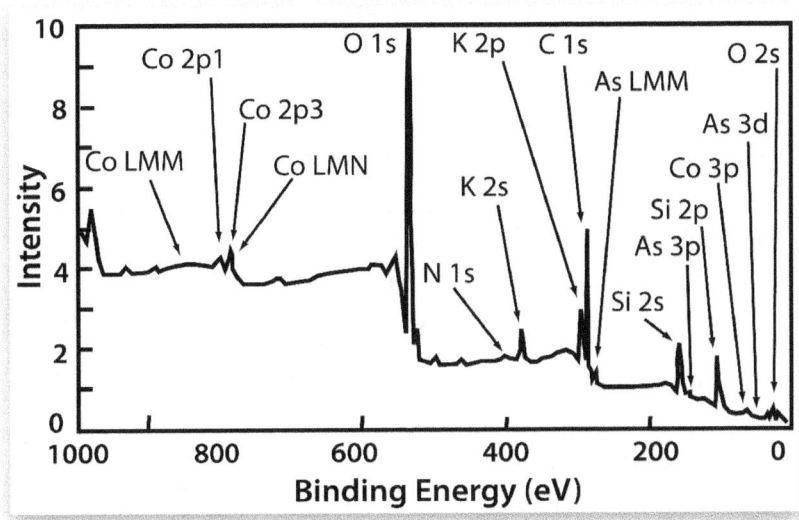

One example of such problem was shown in the Janssens XPS spectrum presented before (Exercise 4.1). Another example is presented in the accompanying figure (adapted from Altavilla et al.) produced by an XPS study of glassy pigments which shows the Co LMM, LMN and As LMM Auger peaks in addition to the XPS spectrum; the nomenclature used for the Auger electrons is based on the shells involved in the transition.

It must be mentioned that Auger electrons obtained through an XPS apparatus appear with the wrong binding energy. This is because they are taken as photoelectrons by the data analysis system. Auger spectroscopy will be discussed at length in the following chapter.

Chemical shifts

The possibility of determining accurate electron binding energies allows XPS to quantify small variations of binding energy due to, for instance, the binding of the atom to another atom (e.g. such as in the formation of compounds). Such variations are known as *chemical shifts* and produce small changes in the location of the peak positions; in oxidation, for instance, atoms lose electrons which in turn increase the binding energy of the photoelectron. XPS is specially suited for detecting the chemical shifts due to the fact that, being a one-electron process, the emitted electrons have a very small energy dispersion, especially in comparison to other processes, such as Auger emission.

One example is titanium which exhibits very large chemical shifts between different oxidation states. The figure shows the electronic configurations of both neutral Ti and O as well as of TiO_2 with the titanium atom losing 4 electrons and the two oxygen atoms receiving them. The lower panel shows the spectrum from a pure Ti sample (Ti^0) compared with that of titanium dioxide (Ti^{4+}); as it can be ob-

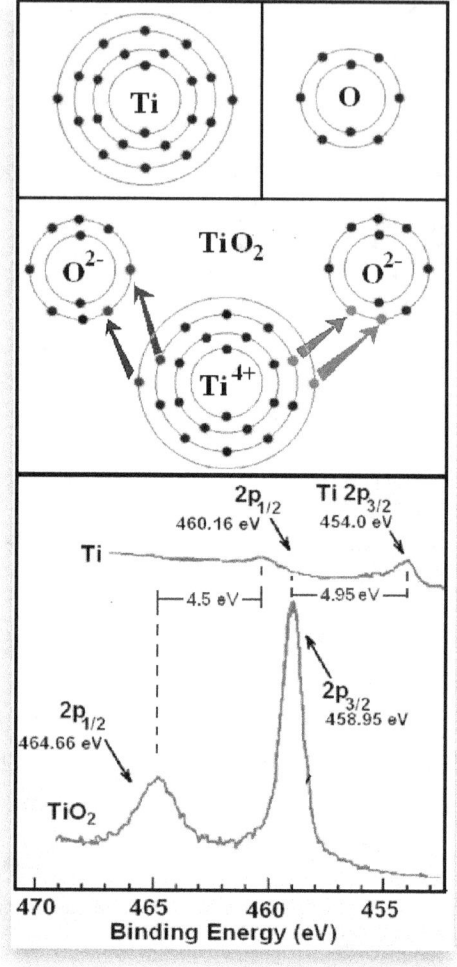

served, the energy shifts are 4.5 eV for the 2p1/2 and 4.95 eV for the 2p3/2.

A more quantitative example of the change in binding energy due to oxidation is illustrated in the XPS spectra of fluorine attached to a silicon surface forming layers of SiF_1, SiF_2 and SiF_3. The left panel of the figure shows the spectra of SiF_x showing the shifted peaks, and the right panel presents the energy shift for four oxidation states. Again, oxidation (i.e. the removal of a valence electron) increases the binding energy; reduction, the addition of an electron, decreases the binding energy.

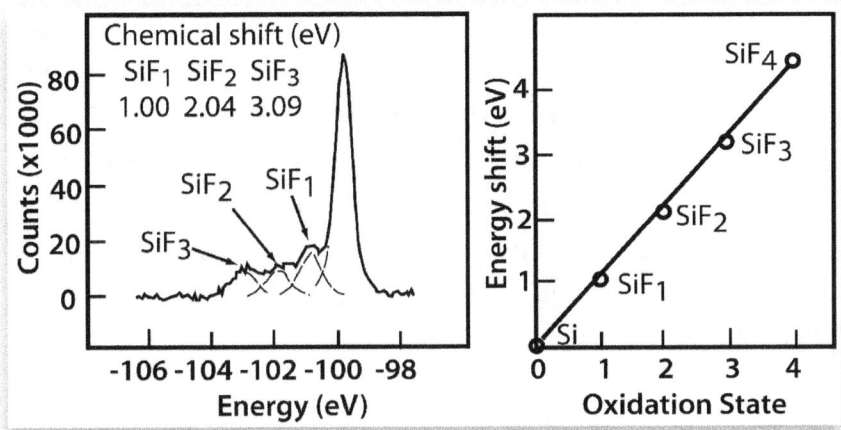

The ability of XPS to resolve chemical shifts depends mainly on the magnitude of the shift, the sharpness of the incident radiation as well as on the energy resolution of the electron detector. Samples with various components that overlap their photoelectron binding energies produce broad XPS peaks, which cannot be resolved for chemical shifts, which are usually in the range of a fraction of an eV to about 10 eV.

Exercise 4.2 Pauling charge model of chemical shift

The determination of the chemical shifts is a difficult problem in general, but in some cases phenomenological theories, such as the Pauling charge model, can be very useful. Simply stated, the theory tracks the number of bonds being replaced in a given structure as atoms bind with another one, and assigns a given energy change per charge replaced. In summary, the change in binding energy is related to the change in the charge on the atom by $\Delta(BE) = k\Delta q$,

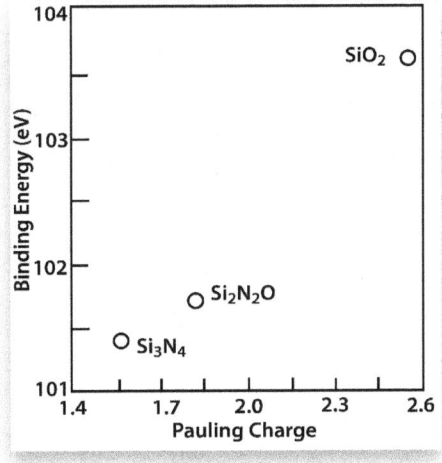

where k is the amount of energy added per every charge change; the calculation of Δq is beyond the scope of these lecture notes, but in general depends on properties of the specific atom of interest.

Using such theory Brow and Pantano studied the shifts of Si 2p in SiO_2, Si_2N_2O and Si_3N_4, and found the binding energies and Pauling charges shown in the plot. The values of Δq are 1.56 for Si_3N_4, 1.81 for Si_2N_2O and 2.54 for SiO_2. Estimate the increase in binding energy due to an increase in charge.

Solution.

Drawing a straight line that best goes through or near the three points, it is possible to obtain the slope of such a curve, which corresponds to the sought answer:
$$k \approx \frac{103.7-101}{2.6-1.45} = 2.35 \ eV \text{ per}$$
added charge. A more professional way is to use a least squares method to determine the slope (and the y-intercept which will be needed in solving Problem 4.5).

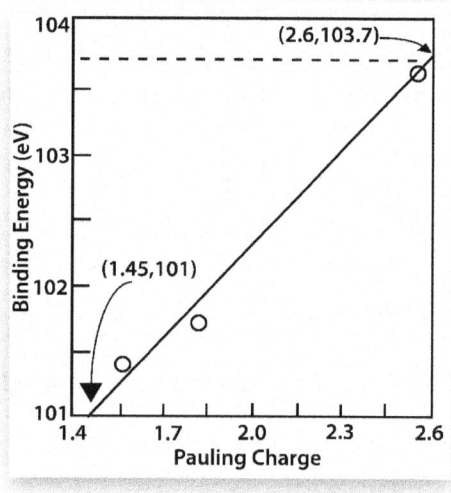

XPS Sampling Depth

The XPS process is obscured by several factors, but most importantly by the absorption properties of the material, both for the incident x rays and the emitted electrons. Although the penetration depth of an x ray is of several μm, the mean free path of an electron is of the order of a few nm, thus strongly limiting the sampling depth of XPS. An accurate description of the absorption of electrons by the medium is a complex process that depends on the kinetic energy of the photoelectron and the properties of the material; in practical terms it is best to describe such absorption in terms of a phenomenological attenuation law known as the Beer-Lambert law. In an XPS analysis the x rays free a number of electrons a depth d below, a fraction of which will move straight to the surface. If the number of electrons that move toward the surface per unit time is I_0, the number of electrons that will reach the surface per unit time, I_S, will be given by

$$I_S = I_0 \, e^{-d/\lambda},$$

where λ is the electron mean free path in the medium. Notice that at depths of $d \approx 3\lambda$ the intensity at the surface is $I_S \approx 0.05 I_0$, i.e. 95% of electrons produced at distances of 3λ will be absorbed; this fact is used to define the "sampling depth" as 3λ.

Luckily, the mean free path, λ, has a behavior that is similar for a

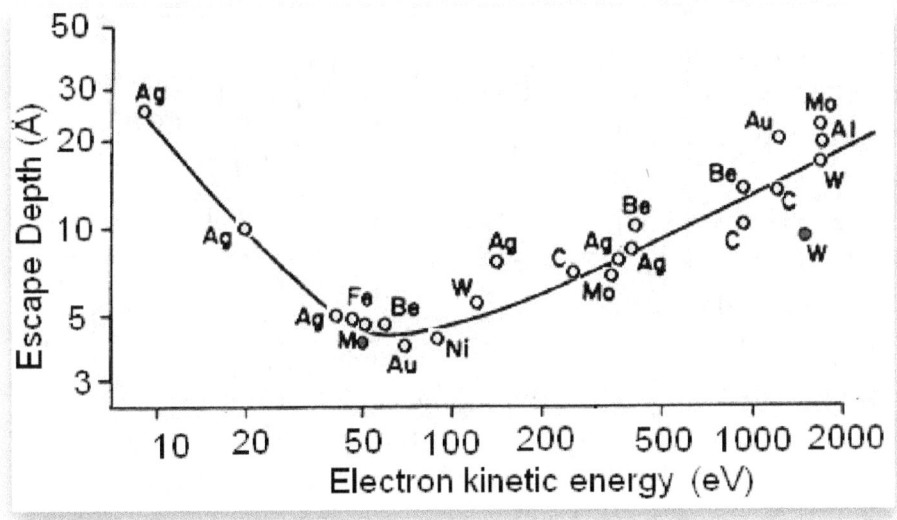

large class of materials and can be characterized in terms of the so-called "Universal Curve" shown in the figure (adapted from the work of Seah and Dench). Experimental measurements in a variety of media show that the sampling depth (or escape depth) of electrons is always in the range of 3 to 30 nm or so. This puts λ in the range of 1 to 10 nm. A complete list of electron inelastic mean free paths can be obtained from the NIST database listed in the Bibliography. In summary XPS can gather information from the top 30 nm of material which corresponds to about 20 atomic layers.

Exercise 4.3 Oxide thickness measurements
Strohmeier in 1990 used the Beer-Lambert law to determine the oxide thicknesses on metals using XPS data. Since both the oxidized and non-oxidized metals produce photoelectrons, it was found that the thickness d of an oxide layer is given by

$$d = \lambda_o sin\theta \; Ln\left[\frac{N_m \lambda_m I_o}{N_o \lambda_o I_m} + 1\right]$$

where λ_o and λ_m refer to the inelastic mean free paths, I_o and I_m to the peak intensities and N_o and N_m to the volume densities of the oxide and the metal, respectively, θ is the photoelectron take-off angle, and Ln represents the natural logarithm function.

As an example consider an Al sample with a top layer of Al_2O_3 studied with XPS producing the peaks shown in the figure (adapted from the X Ray Photoelectron Spectroscopy Reference Pages). The peaks shown correspond to 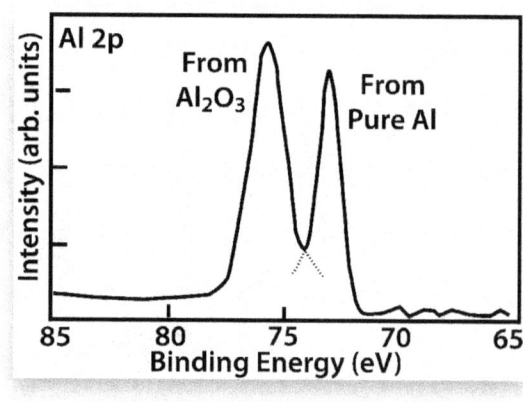 the 2p level of Al from pure metal (at 72.7 eV) and from Al_2O_3 (75.9 eV), and their intensities are I_o = 65 and I_m = 35 (in arbitrary units). If N_m/N_o = 1.5/1.0 and λ_o = 28 nm and λ_m =26 nm, and the photoelectrons were captured normally to the metal surface, the oxide thickness is

$$d = \lambda_0 sin\theta \; Ln \left[\frac{N_m \lambda_m I_o}{N_o \lambda_o I_m} + 1 \right]$$

$$= (28 \text{ Å})(sin 90) \; Ln \left[\frac{1.5 \times 26 \times 65}{1.0 \times 28 \times 35} + 1 \right] = 35.76 \text{ Å}$$

An oxide thickness calculator implemented in Excel can be found at http://sprocket.ssw.uwo.ca/xpsfiles/.

Plasmon peaks

Other effects that modify the XPS spectra are the so called plasmon peaks, which are produced by the interaction of the photoelectrons with collective excitations of the electrons in the valence band known as *plasmons*.

The plasmon excitations are fluctuations of the electron density produced either by the photoelectron itself while it propagates through the solid (extrinsic plasmon), or by the valence electrons in response to the presence of the core hole left by the emitted photoelectron (intrinsic plasmon). Interactions of the photoelectron with the plasmons reduce the kinetic energy of the electron making it appear with a larger binding energy in the XPS spectrum. Due to the repetitive spatial pattern of the plasmon waves, the interactions between the photoelectrons and the plasmons can occur more than once producing several "plasmon peaks" in the XPS spectrum at regular energy intervals. The figure (adapted from Alexander et al.) shows the Al 2s and Al 2p XPS peaks along with several plasmon peaks.

The energy intervals of the plasmon "satellite" peaks are dictated by the material and can provide useful chemical information. Plasmons are noticeable mainly in Al, Mg and Si and are absent in non-metals.

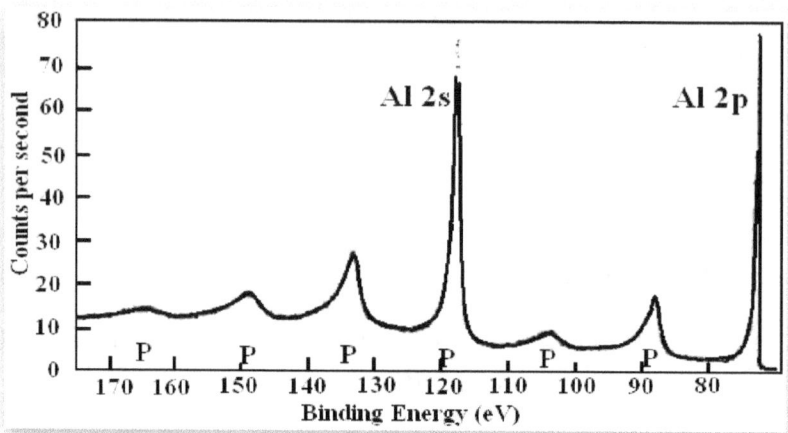

A similar effect is produced when the photoelectron excites a valence electron into a higher energy level (i.e. it "shakes up" the electron). As before, the electron kinetic energy is reduced and the electron is counted in the spectra as if it had a larger binding energy. Due to the narrowness of the valence energy band, these shake ups tend to reduce the kinetic energy by a fixed amount thus producing regular distinct "shake up" peaks at binding energies a few eV larger than the real one.

A related effect occurs when the photoelectron kicks out a valence electron ("shake off"). Since such inelastic collisions can reduce the electron kinetic energy by a wide range of energies, in the end the electrons appear at apparent larger binding energies –not a series of narrow peaks as before— but in a broad "shake off" peak.

Static charging

Yet another phenomenon that can affect the XPS spectrum arises due to the possible build up of positive charge on the surface of non-conducting samples. When thousands of photoelectrons are emitted and the sample is non-conducting, an excess of positive charge is created on the surface of the sample which, in turn, creates a surface electric field that will reduce the kinetic energy of the ejected electrons shifting the XPS peaks to higher binding energies. Fortunately, such effect can be eliminated by shifting the spectrum according to a well known reference point, such as the "adventitious carbon line" (C 1s line of 284.8 eV). Other -hardware based— approaches are depositing a thin layer of gold over the sample, or using low energy electrons from an electron gun during the x ray irradiation to compensate the charge deficiency.

When the sample charges negatively (due e.g. to electron backscattering), the kinetic energy of the photoelectrons increases due to electric repulsion and thus the electrons are counted as if they had a lower binding energy. In such cases the XPS analysis is impossible.

XPS Spectrometer

XPS is usually performed in commercial units or using synchrotron light sources. The picture shows the Perkin Elmer Φ560 XPS/AES/SIMS of the University of Texas at El Paso with the author (López) and students Enrique Ramírez Homs and Tony Rodríguez.

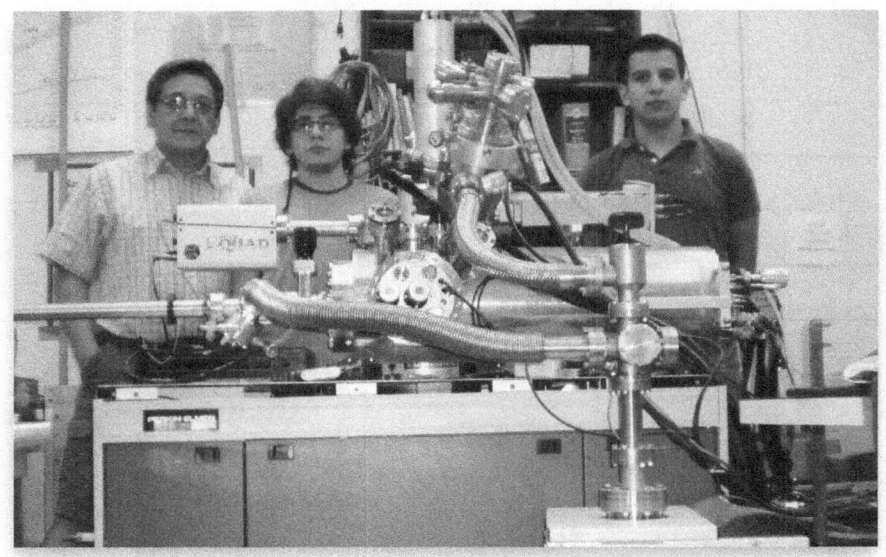

As shown in the schematic, the main components of an XPS system are a source of x rays, an ultra-high vacuum (UHV) chamber with pumps, an electron collection lens and an electron energy analyzer.

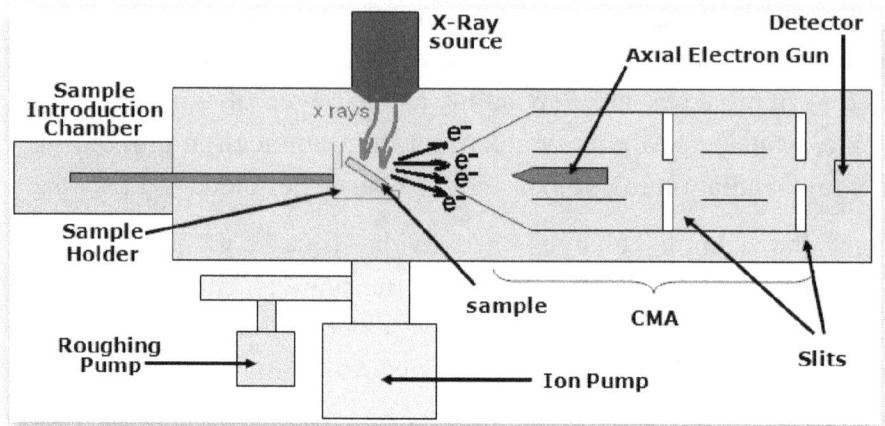

Other elements include an introduction chamber, sample mounts and a data collection system. A useful additional item is an ion gun used to bombard Ar or Ne gas ions to remove surface contaminants from the sample.

The diagram shows the main components of an XPS system such as the Φ560 which operates with either Al Kα (1486.6 eV) or Mg Kα (1253.6 eV) x rays at a pressure smaller than 10^{-9} torr.

The next diagram illustrates the operation of the double-pass cylindrical mirror analyzer (CMA) which is used to collect and analyze the electrons. The CMA is composed of two concentric metallic cylinders maintained at different voltages. By using a combination of electric fields, the CMA captures and guides electrons of a given energy toward the detector while deflecting the

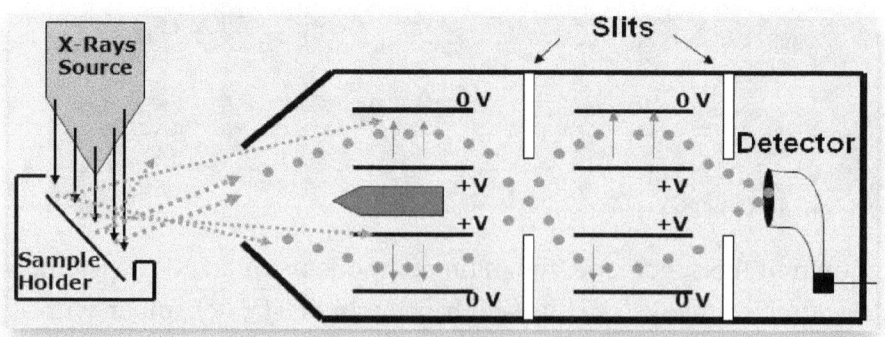

rest towards the shield. By repeating the process at different values of the fields, electrons of different energies are captured and counted. The energy binning determines the resolution of the XPS peaks, which is normally an eV or less; analyses of oxidation states require finer resolutions. An alternative to the CMA is the concentric hemispherical analyzer (CHA) or hemispherical deflection analyzer (HDA); the mechanism being basically the same as in the CMA geometry.

Other considerations

Due to lack of space many other factors that affect XPS are not described in detail; here we mention a few for completeness. One important issue is the energy resolution of the primary x rays, the wider the range of energies of the incoming x ray, the wider the XPS peaks will be. Common widths of primary x rays (e.g. Al Kα and Mg Kα) are between 0.7 eV to 0.8 eV, but can be reduced to 0.4 eV by means of *monochromators* which Bragg reflect X-rays on a crystal but reduce the intensity up to 40%. Since focusing x rays is still an art, beam diameters tend to be wider than electron beams in Auger spectroscopy. In regard to x rays beams used in XPS studies, the beam diameters can be of the order of microns in modern equipment; comparatively speaking, electron beam diameters have sub-micron sizes, e.g. for an AES-equipped system.

Angular resolved XPS (ARXPS) takes advantage of angular incidence of the x rays to fine tune the penetration into the material. For instance, at normal incidence, i.e. at 90°, 95% of the signal intensity emerges from a distance 3λ (three attenuation lengths as defined in the Beer-Lambert law.). In comparison, at 15°, the signal is mostly produced by atoms within a depth of 0.8 λ.

Sputtering samples with ions (mostly Ar) with energies of, say, 1 to 5 keV helps to eliminate layers of surface atoms to make underlying layers accessible for XPS analysis. With "normal" effort, layers up to 1 to 2 μm can become accessible, while prolonged sputtering (of hours) can drill holes and reach the sample substrate. An ever-present problem is the recombination of the eliminated atoms with the underlying layers which can yield to erroneous interpretation of the depth profiles obtained.

Further reading

Interested readers can find a list of more advanced textbooks in the section of general sources in the Bibliography along with a list of sites which provide atomic binding energies, XPS spectra, electron mean free paths, among other data.

Problems

Problem 4.1

The spectra shown was obtained by irradiating Al foil with Mg x rays (1253.6 eV), use the binding energies listed in Table A2.2 in Appendix 2 to identify the origin of the peaks signaled with a line. Notice that the energy resolution of the spectrometer was 1 eV. Caution: before you start finding unexpected elements such as vanadium, manganese, potassium or indium in a simple aluminum foil like the one you may use to wrap your burritos, think about what elements or contaminants could possibly be in or on the foil.

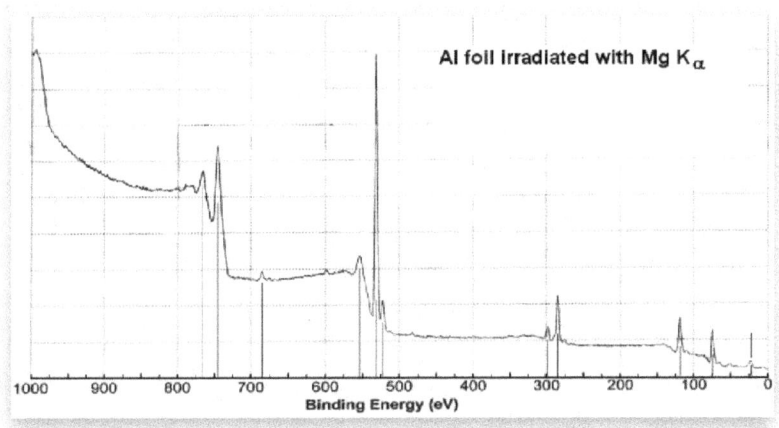

Problem 4.2

A clumsy student obtained the attached XPS spectra but forgot to identify the peaks. Please look the energies of the peaks, their relative areas and the binding energy table A2.2 of Appendix 2 to determine the element(s) which produced the peaks and the energy levels to which the two peaks correspond. Please explain your reasoning.

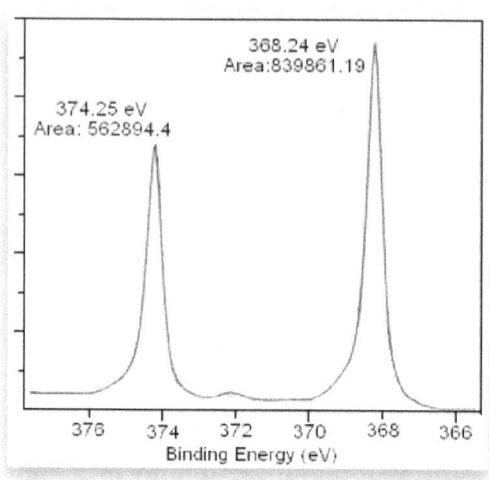

Problem 4.3

The XPS spectrum shown was obtained irradiating Mg K_α x rays on a Pd sample.

A) Determine the binding energy of the electrons that were collected on the five peaks shown.

B) Identify the energy levels of the peaks.

Problem adapted from data of Queen Mary University of London.

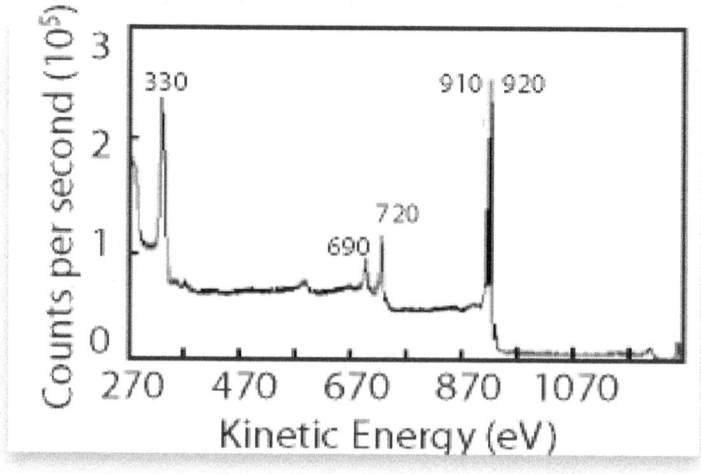

Problem 4.4

You are about to perform a rare XPS study of NaCl using oxygen K_α x rays, what will be the kinetic energies of the 1s, 2s, 2p and 3s electrons of Na?

Problem 4.5

Estimate the oxide thickness of the Al sample studied by Alexander et al. that produced the XPS spectrum shown in the graph. As explained in the article, $N_m/N_o = 1.6/1.0$, $\lambda_o = 2.92$ nm, $\lambda_m = 2.39$ nm, and the photoelectrons were captured in the direction normal to the metal surface. Feel free to use the oxide thickness calculator found at http://sprocket.ssw.uwo.ca/xpsfiles/.

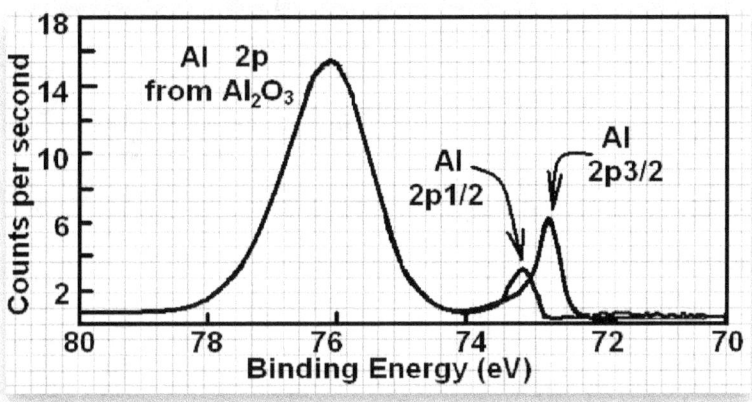

Problem 4.6

Use the information of Exercise 4.2 to estimate the binding energy of pure Si.

Problem 4.7

A sample was analyzed using Mg Kα x rays (energy 1253.6 eV). Use the table of atomic binding energies in Appendix 2 to identify the spectral lines signaled by the lines.

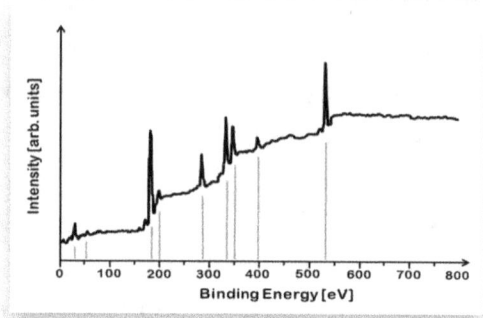

Problem 4.8

In the figure we observe the XPS spectrum taken using Mg Kα x rays. The binding energy is indicated. Identify the element(s).

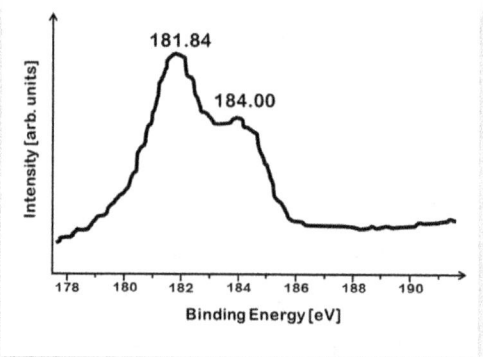

Problem 4.9

The spectrum in the figure was obtained using Mg Kα x rays. The sample is Ge\Si(001).
A) Determine the binding energy from the kinetic energy.
B) Identify the elemental levels.

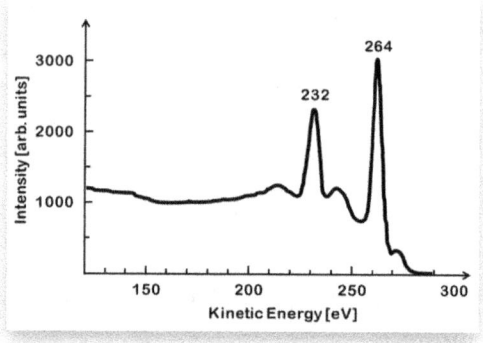

Problem 4.10

In XPS, other sources can be used for analysis. Say that we use Ti Kα with energy of 4510 eV, what would be the kinetic energies of Ca 1s, 2s, 2p, 3s, and 3p, using the binding energies from the table in the Appendix 2?

Problem 4.11

The intensity profile of the figure was obtained to estimate the thickness of SiO_2 on Si and it shows, from high to low BE, the Si 2p1/2 and Si 2p3/2 peaks of the silicon oxide; the data was collected at $70°$ with respect to the

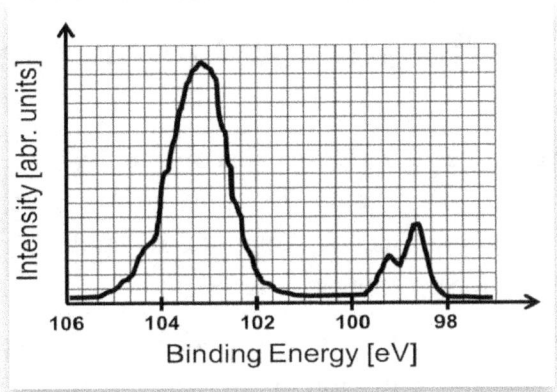

sample normal. In lieu of tabular information about the density of each material and of the inelastic mean free path of the bare silicon, assume that you have measured an infinitely thick sample of SiO_2 on Si, and your measurement yielded a ratio of 0.75. Assume that the λ_{SiO2} = 2.7nm. Using the grid accompanying the figure, estimate the thickness of the SiO_2.

Chapter 5: Auger Spectroscopy

The Auger effect was discovered by Lise Meitner in 1923 and Pierre Auger in 1925, but it was not until 1953 that was implemented as the technique to study surfaces known as Auger Electron Spectroscopy (AES).

The Auger effect is a sequence of three steps as indicated in the figure: the emission of a core electron induced by an external electron beam or an x ray beam, followed by an electronic de-excitation transition into the hole left by the emitted electron, and finalized by the emission of a second (Auger) electron using the de-excitation transition energy. Information about the energy levels of the material can be obtained from the kinetic energy of the Auger electron.

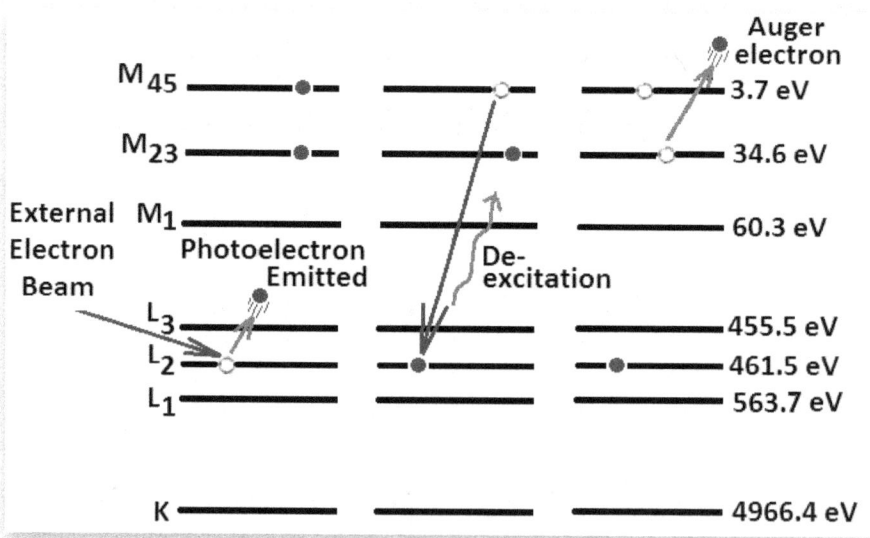

As an example, consider the previous figure which shows the actual values of the energy levels of titanium. The kinetic energy of the Auger electron is the difference between the energy released in the de-excitation and the binding energy of the Auger electron; using the information of the figure:

$$KE = (E_{L2} - E_{M4}) - E_{M2} - \Phi = (461.5 - 3.7) - 34.6 - \Phi$$
$$= 423.2 \text{ eV} - \Phi,$$

where Φ is the work function of the material.

Notice that KE depends solely on the energy levels of the material and is independent of the energy of the incident electron beam. In general AES transitions are described by the three energy levels involved in the Auger process. In the previous example the transition would be labeled as an $L_2M_4M_2$, where L_2 indicates the energy level of the emitted photoelectron, M_4 the level left by the electron in the de-excitation, and M_2 the level

N	l	j	XPS	X-Ray
1	0	1/2	$1s_{1/2}$	K
2	0	1/2	$2s_{1/2}$	L_1
2	1	1/2	$2p_{1/2}$	L_2
2	1	3/2	$2p_{3/2}$	L_3
3	0	1/2	$3s_{1/2}$	M_1
3	1	1/2	$3p_{1/2}$	M_2
3	1	3/2	$3p_{3/2}$	M_3
3	2	3/2	$3d_{3/2}$	M_4
3	2	5/2	$3d_{5/2}$	M_5

from which the Auger electron was emitted. The energy levels are labeled according to the x ray notation shown in the table.

The fact that the *same* electron could have been emitted after an emission of an L_1 or L_3 electron resulting with energies of 525.4 eV or 417.2 eV, respectively, produces a series of individual peaks in the AES spectrum. More often than not, these peaks overlap, making them appear as a single broader peak generically labeled as LMM; this effect is known as *multiplet splitting*. Auger electrons emitted from the valence band are denoted by V.

An important difference with x ray fluorescence is that the Auger effect is *radiationless*, i.e. the intermediate de-excitation process does not emit an x ray. The fact that such energy is used internally to emit the Auger electron allows for the de-excitation transition to take place between levels that otherwise would be forbidden under the $\Delta l = \pm 1$ rule strictly respected by XRF.

[A similar effect is also observed in the so-called *internal conversion*, in which an excited nucleus de-excites –not by emitting a gamma ray but— by ejecting an atomic from a core orbital; such process also fails to respect the $\Delta l = \pm 1$ rule.]

Exercise 5.1 illustrates the possible set of values of the kinetic energy of the KLM Auger electrons in silver, including one such formerly forbidden transitions from 2s to 1s (L_I to K).

In general, the kinetic energy of a, say, KLM Auger electron emitted from a material with a work function Φ (usually of the order of a few eV) is:

$$KE = (E_K - E_{L*}) - E_{M*} - \Phi.$$

The asterisks indicate that both E_{L*} and E_{M*} corresponds to energy levels in an *ionized* atom, i.e. in an atom with one missing electron. To a good approximation, ionized atoms have energy levels

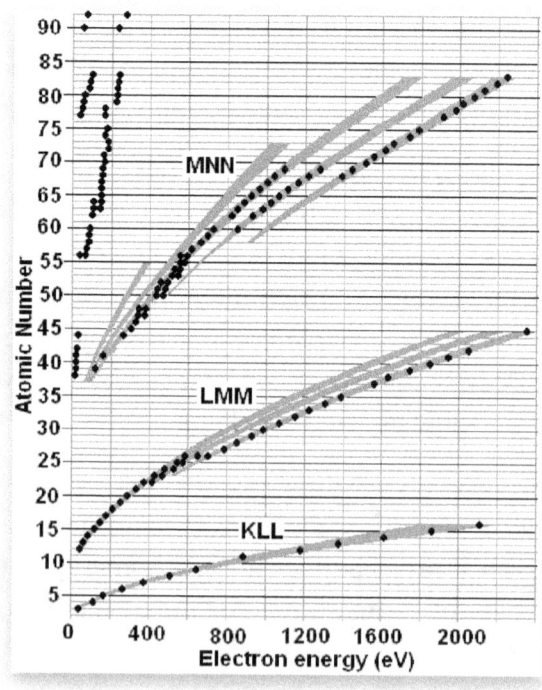

intermediate to those of neutral atoms with Z electrons and with Z+1 electrons, namely, $E_{L^*} = \frac{1}{2} [E_L(Z) + E_L(Z+1)]$ and $E_{M^*} = \frac{1}{2} [E_M(Z) + E_M(Z+1)]$. The chart (adapted from Childs et al, 1995) shows the energy of the most common Auger peaks (dots) and less intense peaks (gray bands) in the KLL, LMM, MNN and higher transitions for elements above cesium. Tables of Auger energies can be found in the Lawrence Berkeley National Laboratory web site http://xdb.lbl.gov/Section1/Sec_1-4.html.

AES Spectrum

The Auger spectrum is a histogram of the number of electrons (ordinate) captured at a given kinetic energy (abscissa), unfortunately not all captured electrons are unscattered Auger electrons and the sought peaks usually lie buried in the background.

The background has three main components: *secondary electrons*, primary backscattered electrons and inelastically scattered Auger electrons. *Secondary electrons* are valence electrons freed with low energy when interacting with the primary beam, they have a power law behavior. Primary backscattered electrons are electrons that penetrate deep into the sample, and come back out the surface after multiple interactions; the behavior has an exponential shape. Inelastically scattered Auger electrons are Auger electrons originated deep into the sample that underwent an energy loss event before emerging out the surface.

The effect of this background varies according to the energy and angle of incidence of the electron beam, but in general tends to produce a hump around the Auger energy and, along with the multiplet splitting, end up producing broad Auger peaks.

Because of this, it is necessary to use a three-step identification process of the Auger peaks as illustrated in the figure: capture of all electrons at all energies N(E), amplification of peaks N(E)×10, and plot of derivative dN(E)/dE. The use of the derivative eliminates the effect of the background leaving a sharp clear signal of the Au-

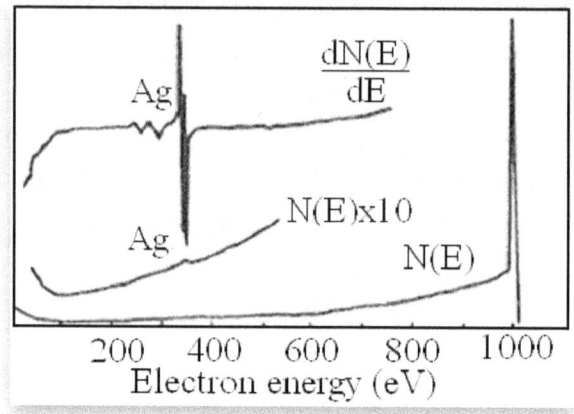

ger effect; by convention, the peak position is taken as the high energy minimum. Another method commonly used to eliminate the background and identify Auger peaks is to plot $d(E \times N(E))/dE$ versus E.

As with XPS, the relative intensities of the Auger peaks depend on the probability of the levels to be ionized by the electron beam; this in turn depends on the atomic number and the electron beam energy. The figure shows the sensitivity for a number of elements and for beam energies of 3 keV and 5 keV.

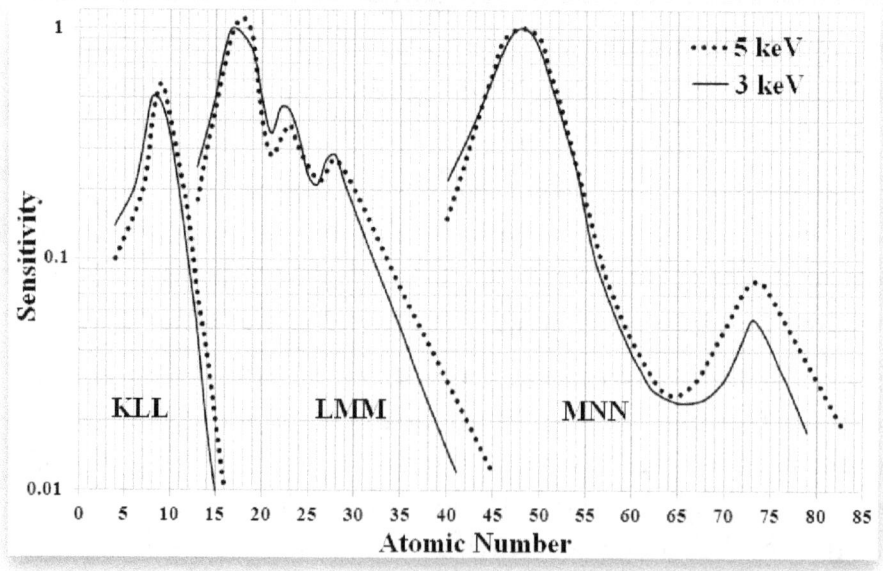

Exercise 5.1

Use the values of the table to determine approximate values of the kinetic energies of all possible KLM Auger transitions in Ag. Neglect changes in energy due to ionization and the small work function ($\approx 4 \pm 1$eV).

Binding energies for silver					
Label	Orbital	Energy(eV)	Label	Orbital	Energy(eV)
K	1s	25514	M_{III}	3p3/2	573
L_I	2s	3806	M_{IV}	3d3/2	374
L_{II}	2p1/2	3524	M_V	3d5/2	368.3
L_{III}	2p3/2	3351	N_I	4s	97
M_I	3s	719	N_{II}	4p1/2	63.7
M_{II}	3p1/2	603.8	N_{III}	4p3/2	58.3

Solution

$$KE = (E_K - E_{L*}) - E_{M*} - \Phi \approx KE = (E_K - E_L) - E_M$$

- KL_IM_I: KE = (25514 – 3806) – 719 = 20989 eV
- KL_IM_{II}: KE =(25514–3806)–603.8= 21104.2 eV
- KL_IM_{III}: KE = (25514 – 3806) – 573 = 21135 eV
- $KL_{II}M_I$: KE = (25514 – 3524) – 719 = 21271 eV
- $KL_{II}M_{II}$: KE = (25514 – 3524) – 603.8 = 21386.2 eV
- $KL_{II}M_{III}$: KE = (25514 – 3524) – 573 = 21417 eV
- $KL_{III}M_I$: KE = (25514 – 3351) – 719 = 21444 eV
- $KL_{III}M_{II}$: KE = (25514 – 3351) – 603.8 = 21559.2 eV
- $KL_{III}M_{III}$: KE = (25514 – 3351) – 573 = 21590 eV

Concentrations

As in XPS, the atomic concentrations can be deduced from AES spectrum due to the fact that the Auger peaks are proportional to elemental concentrations. The task, however, is complicated by many factors that influence peak heights, namely, beam energy, sample orientation, energy resolution and acceptance angle of the analyzer, chemical states of elements, sample heterogeneity, etc.

The first step in finding the concentration of an element is to determine the background curve under the elemental peak followed by the determination of the intensity of the peak (I_x). Since each element has a

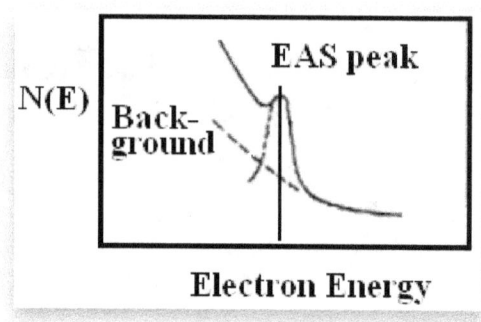

different response to the Auger effect, each peak must be normalized by the elemental sensitivity (S_x) dividing the peak intensity by the sensitivity, I_x/S_x. Repeating these steps for all the elements in the sample it is possible to obtain the concentrations as percentages of the sum of all elements by means of

$$C_x = \frac{I_x/S_x}{\sum_{y=\,All\,elements} I_y/S_y}.$$

To obtain exact concentrations, however, it is necessary to compare to a calibrated sample such as those that can be purchased from the National Institute of Standards and Technology, NIST.

Spectrum effects

The use of an electron beam yields a vast production of *secondary electrons* and *backscattered electrons*. The secondary electrons are low energy electrons (<50 eV) liberated by the beam of electrons through inelastic collisions, while the backscattered electrons are beam electrons that bounce back at large angles maintaining most of their kinetic energy. These electrons constitute the main component of the background.

An effect related to the background electrons is the *charging effect*. Since AES uses an electron beam, it cannot be performed in non-conducting samples without discharging the sample during electron irradiation, but even in conducting samples it is possible to produce an accumulation of charge during the AES irradiation. Charging results when the number of electrons leaving the sample is not properly replaced with incident electrons; this can lead to positive or negative charging. Unfortunately, the methods used in XPS, such as the use of a second electron gun or

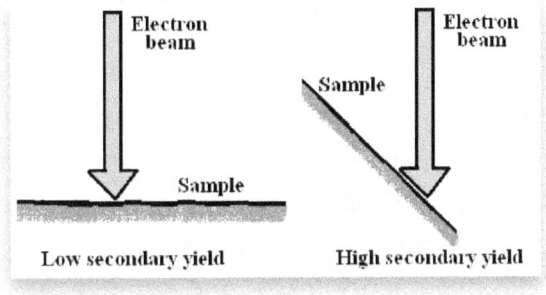

of metallic powder, distort the AES signal. In the case of AES, other methods are to be used to maintain the charge balance of the sample, e.g. conducting clamping.

A trick of the trade that appears to work is to reduce the number of incident electrons (and thus the emitted electrons) by directing the electron beam at a glancing angle of ~10° while carefully controlling the bombarding energy. Other possibility is to use a neutralizing gas, such as Al_2O_3 in Ar at ~10^{-4} Torr, or O_2 at ~10^{-8} Torr, which have been shown to stop charging for hours. The figure (adapted from Guo et al.) shows the AES spectra of SiO_2 in UHV (top) after 12 hours of exposition to the AES electron beam, and in an O_2 atmosphere at 5×10^{-8} Torr (bottom) after 6 minutes (bottom)

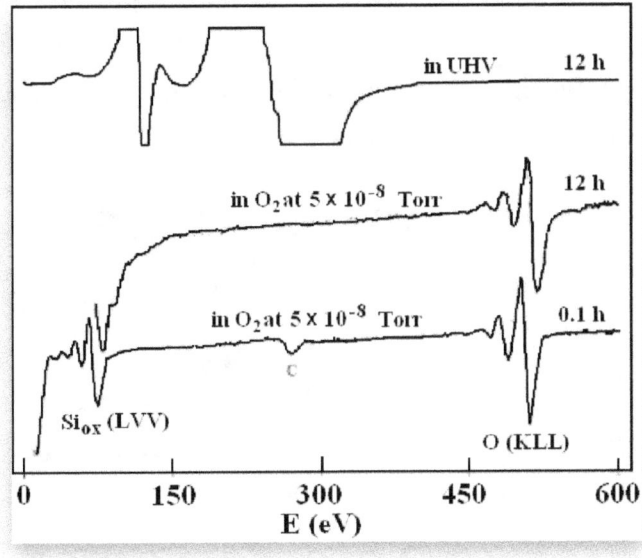

and 12 hours (middle) of irradiation; the deterioration of the spectra in UHV is evident.

The *plasmon* effect discussed in XPS is also present in AES. Emitted electrons can undergo the characteristic energy losses through multiple scattering events especially with the collective electron density oscillations called plasmons. At a difference from XPS, the energy losses due to interactions with the plasmon are of the same order as the Auger peaks and can make the background surpass the peaks of interest; in this case it is difficult to extract the Auger information from the background. Other effect than can modify the Auger spectrum are the so-called *"satellite peaks"* which are due to multiple ionization of a single atom.

SEM imaging

A plus of AES is its ability to produce elemental maps which can be combined with secondary electron images. By scanning the electron beam over a surface, it is possible to capture a large number of secondary and backscattered electrons and use this information to produce a scanning image. Combining such micrograph with the information obtained from AES it is possible to obtain elemental maps of regions of the sample. The figure shows a SEM micrograph (right) of about 10 μm × 10 μm of a gold foil along with the corresponding AES spectrum (left) showing some prominent Au peaks. This ability of AES, to produce

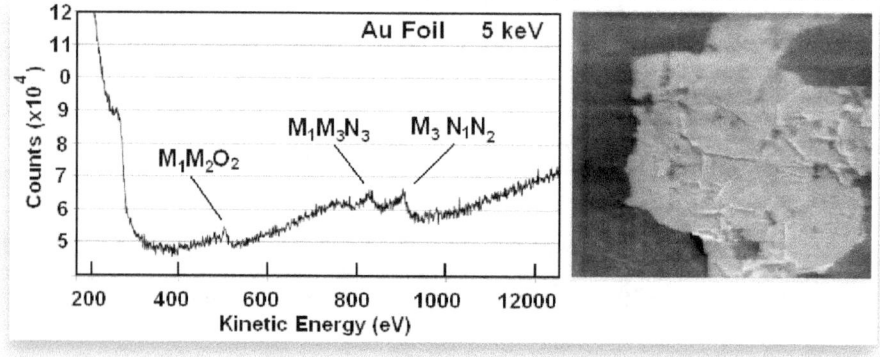

elemental maps, is available, mainly, due to the small spot size (high resolution) of the electron beams which have diameters that range from hundreds of μm to tens of nm.

Comparison to XPS and XRF

A major difference between the photoelectric effect, x ray fluorescence and the Auger effect is that first one involves only one atomic electron, the second two and the third one three; in fact XRF and Auger emission always take place after a photoelectron has been emitted and they are competing and mutually exclusive effects.

In practice, after the core electron has been emitted, the atom de-excitation can result on the production of an escaping x ray —in a case of x ray fluorescence— or in the emission of an Auger

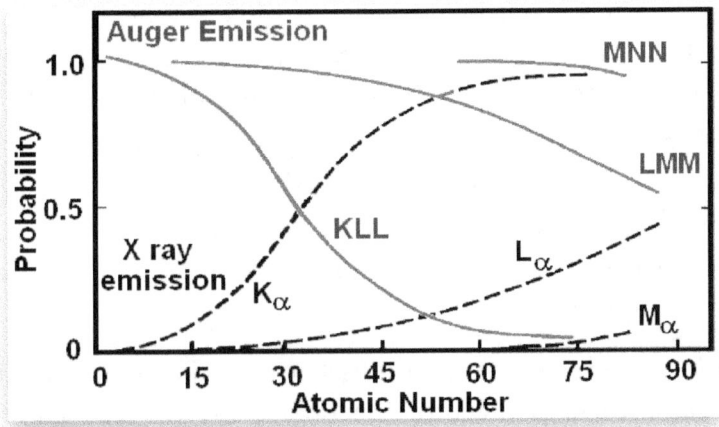

electron. The relative probability of these competing processes is shown in the figure, and it indicates that Auger emission is more likely than x ray fluorescence for lighter elements, but the situation reverses for elements with higher numbers of electrons.

Comparing AES to XPS one finds several differences. For starters, AES is more efficiently initiated with an electron beam, AES started with x rays (of 1 to 2 keV) is known as XAES. In AES the analyzed electrons are not core electrons (as in XPS) but higher

83

energy (less bound) electrons. Another difference already mentioned is that the kinetic energy of the AES electrons is independent of the energy of the electron beam; the only requirement is that the electrons in the beam have enough energy to eject a core electron. A consequence of the use of an electron beam as the primary radiation is that the depth of the analysis in AES is reduced to the 10 to 100 nm range, much smaller than that of XPS which reaches up to 100 μm. Finally, AES can be more easily carried out on conducting and semiconducting materials, but insulating materials can also be studied by either using a conductive support or thinning the sample down to micron or submicron thicknesses. Samples that cannot sustain electron irradiation cannot be studied.

Continuing with the differences, the Auger spot size is smaller than that of XPS, but AES is not as well suited to study chemical shifts as XPS. Using x rays, XPS is less destructive than the electron beam of AES. A clear advantage of AES is its imaging capabilities discussed before. Yet another difference between XPS and AES is that real XPS peaks in an XPS spectrum do not change if the x ray source is changed (e.g. from Mg Kα to Al Kα), while the fake AES peaks move by the energy difference of the x rays (233 eV for the switch of Mg Kα to Al Kα); this allows the two types of peaks to be easily distinguished.

On the other hand, some similarities between the two methods is that both can measure concentrations up to parts per thousand (ppt, or 0.1%), quantification is not easy and requires the use of calibrated samples, techniques require of UHV conditions, are non-destructive, require small amounts of material for an analysis, and do not require specialized sample preparation. A more graphical comparison of the surface techniques is presented in Appendix 3. Estimated times to obtain an AES survey spectrum from a sample vary from 1 to 5 min, much like that of XPS, although detailed high-resolution maps require longer scanning times.

AES Spectrometer

The main components of any AES system are an electron beam gun, an UHV chamber, an electron collection lens and an electron energy analyzer. Due to the similarity with the equipment needed for XPS, these two techniques can be merged into a single apparatus, see e.g. the Perkin Elmer Φ560 XPS/AES/SIMS shown in the previous chapter. Another element usually included in AES spectrometers is an ion gun used for surface sputtering of inert gases, such as Ar or Ne, either to perform Secondary Ion Mass Spectroscopy (SIMS) or to simply eliminate atomic layers and study the subsurface of a sample.

The schematics of an AES spectrometer are shown in the figure; the sample is bombarded with electrons from an electron gun, and the emitted Auger electrons are analyzed in a CMA or CHA (also known as hemispherical mirror analyzer, HMA). As with XPS, AES requires the use of an UHV chamber with pressures smaller than 10^{-9} Torr to increase the mean free path of electrons of up to 2 keV and to reduce the influence of residual elements deposited on the surface. Higher chamber pressures can be used as long as the pumps used can handle it and the signal of interest is not adversely affected by either contamination and/or reduced electron yield. Note that for a pressure of about 10^{-6} Torr, a monolayer coverage of nitrogen can occur in about two seconds (see Problem A1.2.)

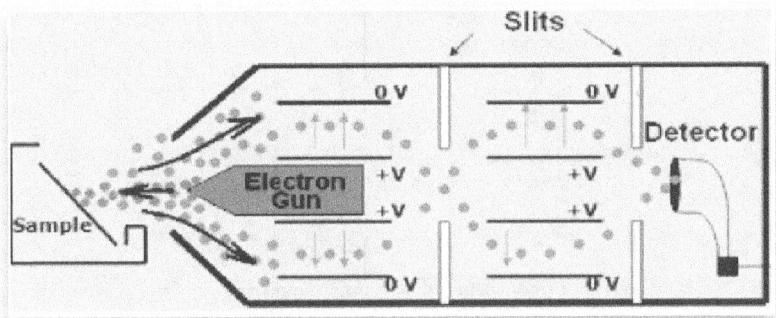

To have sufficient energy for the Auger process and ionize K, L or M level electrons, the electron beam should have energies of about

5 times the binding energy; in practice primary energies of 3 to 10 keV are used. The beam current is normally between nA and μA and it is produced by a tungsten filament which, unfortunately, must be replaced every 100 hr of use; other sources (lanthanum hexaboride) produce intensities higher than tungsten but are more expensive. Normal spot sizes of the order of a micron can be achieved with 3 to 10 keV beams. The electron energy analyzer is usually a CMA similar to the one discussed for XPS.

Elastic peak alignment

Because of the many factors that can modify the AES spectrum, a proper identification of the energy peaks requires the use of an energy calibration technique. The usual method is to pick a low beam energy (e.g. 2 keV) and adjust electron gun until the elastic (backscattered) peak is found in the signal at the proper energy. This, in practice, finds the focal spot of the CMA and leaves the beam ready for a correct measurement of peak positions, intensities and peak ratios. The figure shows a typical elastic peak alignment at 2 keV.

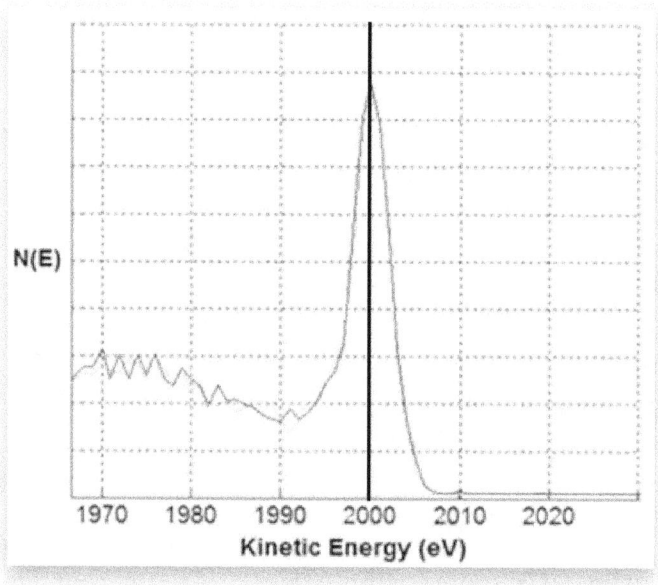

Other considerations

There are many other factors that play a role in AES. The electron beam diameter, for instance, is correlated with the beam current, the smaller the current the thinner the spot size. Beam diameters have a large influence on the number of backscattered electrons produced and, hence, on the SEM images being produced.

Further reading

A sizable list of advanced books on AES is presented in the section of general sources in the Bibliography. Also included are several web sites with information about Auger peaks, and even a concise AES training tutorial.

Problems

Problem 5.1

Determine the number of possible LMM transitions that can take place in Ni, and evaluate the kinetic energy of five cases. The right panel of the figure shows the energy levels, a sample transition and the left panel shows the AES spectrum. Neglect changes in energy level due to ionization and the small work function.

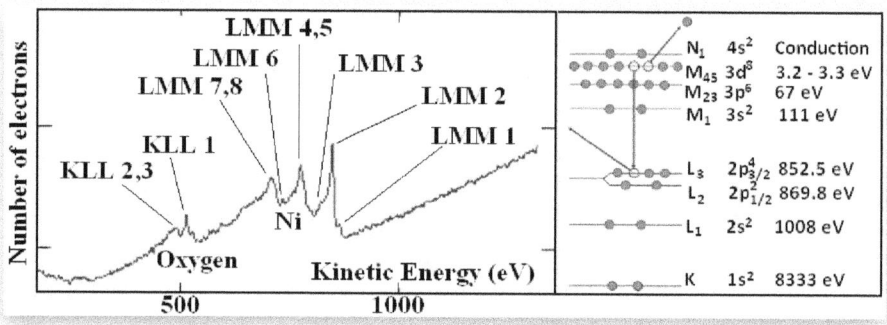

Problem 5.2

Carbon is being bombarded by Al Kα x rays of 1485 eV, if C has the 1s, 2s and 2p levels at binding energies of 290 eV, 20 eV and 15 eV, respectively; sketch the spectrum (number of counts vs. kinetic energy) you expect to see. Include both XPS and AES peaks and identify them using the proper nomenclature for each. Neglect the work function.

Problem 5.3

An Auger peak can be described by a Gaussian curve given by

$$N(E) = \frac{N_0}{\sqrt{2\pi\sigma^2}} e^{-(E-E_0)^2/2\sigma^2}$$

where $N(E)$ is the number of Auger electrons obtained at an energy E, E_0 is the mean energy of the peak, σ is the width of the bell shape curve at half maximum, and N_0 is a normalization constant.
A) Determine total number of electrons counted in an Auger peak.
B) Derive an expression for $dN(E)/dE$, and sketch both $N(E)$ and $dN(E)/dE$ on the same axis.

Problem 5.4
Estimate some LMM Auger transitions in Cu electron binding energies (in eV).

	K	L-I	L-II	L-III	M-I	M-II	M-III
Cu	8979	1096.7	952.3	932.7	122.5	77.3	75.1

Problem 5.5
If we were to use Mg K-alpha radiation (1253.6 eV) to study the oxygen XPS and AES lines, what would be the expected energies of the spectral lines? Use the following binding energies for oxygen 1s: 539 eV, 2s: 31 eV, and 2p: 24 eV.

Appendix 1: Vacuum Technology

To avoid collisions between the incident or emitted x rays or electrons with air molecules, spectrometers must operate in conditions of extreme low pressure, i.e. in what is known as ultra high vacuum (UHV), which is characterized by pressures smaller than 10^{-7} Pa. In this Appendix we briefly review the basic principles behind the operation of ultra high vacuum chambers.

Basic Concepts

The interest in vacuum arose in the 17th century, when engineers realized the difficulties of pumping water out of mines from depths larger than the limit of about 30 feet depth. Galileo was commissioned to investigate the problem and began by investigating the properties of air; he then constructed pistons and a 34 ft tall barometer. Years later Evangelista Torricelli constructed a mercury barometer and also concluded that it was the atmospheric air that forced water to a height of 33.6 ft; he was also the first in producing a sustained vacuum. Blaise Pascal, in 1644, sent a group of mountaineers to the Alps to verify that pressure varies with height. H. McLeod in 1872 figured a variation of a manometer that could measure small pressure variations.

Pressure is the force exerted per unit area on a surface. Solids can only exert pressure due to their weight and this pressure is usually downward in the direction of gravity. Liquids also exert pressure due to their weight, but in this case it can be directed sideways as well as downwards. Gases, in addition to exerting pressure due to their weight can also exert pressure due to the motion of the gas molecules. Pressure due to the weight is, for instance, the atmospheric pressure; in closed containers pressure by weight is negligible.

Most gases obey the *ideal gas law*,

$$pV = \nu RT = NkT,$$

where p is the pressure, V the volume, ν is the number of moles, R the gas constant, N is the number of molecules, k is Boltzmann's constant, and T the temperature. Specifically, the number of moles is $\nu = m/M$ with m being the particle mass and M the molar mass. R is related to the Boltzmann constant, k, and Avogadro's number: $R = k\,N_A$. R can be expressed in different units:

$$R = 8.31451 \text{ J K}^{-1} \text{ mol}^{-1} = 8.20578\times10^{-2} \text{ L atm K}^{-1} \text{ mol}^{-1}$$
$$= 8.31451 \times 10^{-2} \text{L bar K}^{-1} \text{ mol}^{-1} = 8.31451 \text{ Pa m}^3 \text{ K}^{-1} \text{ mol}^{-1}$$
$$= 62.364 \text{ L Torr K}^{-1} \text{ mol}^{-1} = 1.98722 \text{ cal K}^{-1} \text{ mol}^{-1}.$$

k is Boltzmann constant: $k = 1.3806\times10^{-23}$ J/K, and N_A is Avogadro's number: $N_A = 6.022\times10^{23}$ mol^{-1}.

Incompressible fluids in a pipe obey the *Bernoulli's equation* which is a mere statement of conservation of energy. Consider the work done in moving a fluid through a pipe, the fluid is being pushed by a force P_1A_1 on the left (see arrow in diagram)

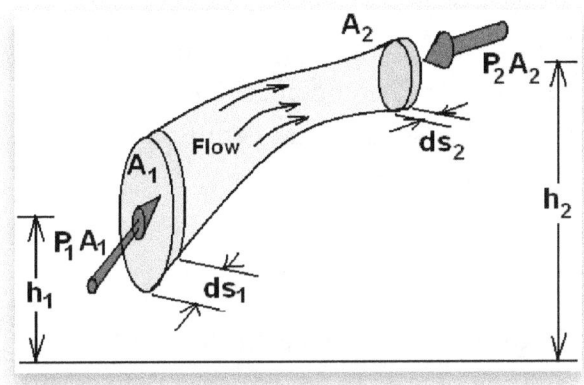

and finds a resistance due to a force P_2A_2 on the right. The work done, force×distance, is equal to the change in kinetic and potential energy of the fluid,

$$dW = P_1A_1ds_2 - P_2A_2ds_2 = d(KE) + d(PE)$$
$$= \tfrac{1}{2}\,mv_1^2 - \tfrac{1}{2}\,mv_2^2 + mgh_1 - mgh_2,$$

since $dV = Ads \Rightarrow P_1 - P_2 = \tfrac{1}{2}\,\rho(v_1^2 - v_2^2) + \rho g(h_1 - h_2),$

which is the Bernouli's equation, also expressed as

$$P_1 + \tfrac{1}{2}\,\rho v_1^2 + \rho gh_1 = P_2 + \tfrac{1}{2}\,\rho v_2^2 + \rho g\,h_2.$$

Exercise A1.1

A) Determine the pressure exerted by a 1 g of H_2 in a 10 liter container at 20 C.

B) Repeat for 1 g of Ar.

C) How do you explain that a lighter gas exerts a larger pressure tan a heavier one?

Solution

A) $p = \dfrac{(1\,g)(83.14\,mbar\,l\,mol^{-1}K^{-1})(293\,K)}{(10\,l)(2\,g\,mol^{-1})}$

$= 1218\,mbar$

B) $p = \dfrac{(1\,g)(83.14\,mbar\,l\,mol^{-1}K^{-1})(293\,K)}{(10\,l)(40\,g\,mol^{-1})}$

$= 60.9\,mbar$

C) 1 g of H_2 contains $[1g/(2g/mol)] \times N_A = N_A/2$ molecules, whereas 1 g of Ar contains $N_A/40$ molecules, i.e. there is a considerable smaller number of molecules of the heavier gas than of the lighter one.

Exercise A1.2

The role of a UHV system is to reduce the pressure, if the volume of the chamber is fixed, how can the pressure be reduced?

Solution

In $pV = \nu RT$, but $\nu = \dfrac{m}{M} = \dfrac{\rho V}{M} = \dfrac{NV}{VM} = \dfrac{N}{M}$, where N is the number of molecules.

Thus $pV = NRT/M$, and p can be reduced by decreasing N or T, or increasing the molar mass M. In practice, N is reduced with the use of vacuum pumps, and T with a cooling system. M is not usually changed as it is not practical to exchange, say, air by a heavy gas to then extract it out of the chamber with a vacuum pump.

For static fluids, such as gases in a chamber, _kinetic theory_ shows that the net velocity of the fluid is zero and only 1/3 of the particle moves on a given direction on average, in that case the change in gravitational potential energy is zero, and the gas pressure is

$$p = \frac{1}{3}\rho\langle v^2 \rangle = \frac{1}{3}nm\langle v^2 \rangle.$$

Where n is the number density (particles per unit volume), and $\langle v^2 \rangle$ is the average velocity related to the temperature by $\langle v^2 \rangle = 3\frac{kT}{m}$, thus the _root-mean-square molecular speed_ is

$$v_{RMS} = \sqrt{\langle v^2 \rangle} = \sqrt{3\frac{kT}{m}} = \sqrt{3\frac{RT}{M}}.$$

Molecule speeds in a gas are distributed according to a Maxwell-Boltzmann distribution with an _average molecular speed_ of $\langle v \rangle = \sqrt{8kT/\pi m} = \sqrt{8RT/\pi M}$; notice that $\langle v \rangle = \sqrt{8/3\pi}\, v_{RMS}$.

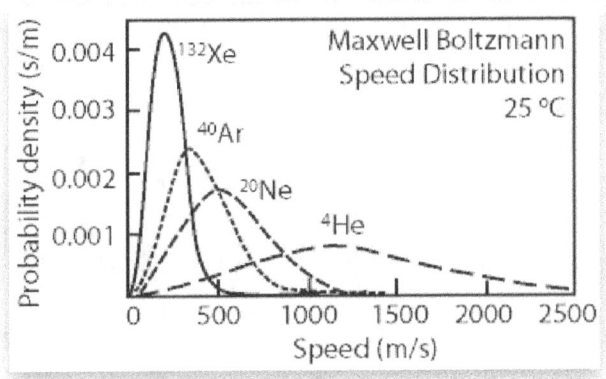

Likewise, assuming that the gas particles are tiny spheres of radius R, with number density ρ= N/V and moving with mean velocity v, it can be shown that the mean time between collisions, i.e. _mean free time_, is

$$t_{Mean} = \frac{1}{4\pi\sqrt{2}R^2 v\rho}.$$

And the mean distance travelled between collisions, i.e. the _mean free path_, is

$$\lambda_{Mean} = vt_{Mean} = \frac{1}{4\pi\sqrt{2}R^2\rho} = \frac{kT}{4\pi\sqrt{2}R^2 p}.$$

Notice that the product $\lambda_{mean} \times p$ = constant. The periodic table of Chapter 1 shows the covalent atomic radii in pm, 10^{-12} m.

Gas particles in the UHV chamber will interact with the sample being studied. It can be shown that the number of gas molecules hitting the sample per unit area per unit of time, i.e. the _flux_ or _rate of arrival_, is given by (see exercise A1.4) as:

$$R = \frac{\rho\langle v\rangle}{4} = \frac{p}{T^{\frac{1}{2}}M^{\frac{1}{2}}}\sqrt{\frac{N_A}{2\pi k}}.$$

In the mks system pressure is measured in N/m^2, i.e. in $kg/s^2 m$, known as _pascal_ (Pa); different fields use different units, the table shows the relationship between various units.

	Pa	Atm	Torr	Bar	PSI
Pascal	1	9.8692×10^{-6}	7.5006×10^{-3}	1.00×10^{-5}	1.4504×10^{-4}
Atmosphere	101,325	1	760	1.0133	14.69597
Torr	133.32	1/760	1	1.333×10^{-3}	1.9337×10^{-2}
Bar	100,000	0.9869	750.06	1	14.5038
PSI	6894.7448	0.0680	51.7147	0.0689	1

The atm is a measure of the mean atmospheric pressure at sea level, a torr is the pressure exerted by 1 mm of Hg, a bar is almost equal to 1 atm, psi is the pressure exerted by a force of one pounds on a square inch. Additionally, in the cgs 1 pa is 10 dynes/cm^2, where a dyne = g cm^2/s^2 and J = 1×10^7 dynes.

Exercise A1.3

A UHV chamber of 11 liters of volume is maintained at a pressure of 7.9×10^{-8} Pa at room temperature (21°C), if it contains mostly air (average molar mass 28 g/mol),

A) how many molecules are there in the chamber?

B) What is the average inter-particle distance between molecules?

C) If an electron beam of circular cross section of 1 mm of diameter travels 2 cm to reach the target, how many air molecules will it encounter in its path?

D) What is the RMS speed of the air molecules in the UHV chamber?

Solution

$$N = nN_A = \frac{pV}{RT} N_A$$
$$= \frac{(7.9 \times 10^{-8} kgm/s^2)(0.011m^3)}{(8.314\,J/mol\,K)(294\,K)} 6.022 \times 10^{23} particles/mol$$
$$= 2.14 \times 10^{11} \text{ particles.}$$

B) Specific volume: $V_{sp} = 1/\rho = V/N = 4\pi R^3/3 \Rightarrow$
   $R = (3V/4\pi N)^{1/3} = [\,3 \times 0.011 \text{ m}^3/(4 \times 3.14159 \times 2.14 \times 10^{11})\,]^{1/3}$
$$= 2.30 \times 10^{-5} \text{ m,}$$
and the interparticle distance is $2R = 4.61 \times 10^{-5}$ m $= 46.1$ μm.

C) Volume swept $V_s = \pi r^2 L$, number of particles in volume swept:
   $N = V_s/V_{sp} = \pi r^2 L/V_{sp} = \pi(0.001/2)^2\, 0.02$ m $/(4\pi R^3/3)$
$$= (0.001/2)^2\, 0.02 \text{ m }/[4(2.30 \times 10^{-5} \text{ m})^3/3] = 308,210 \text{ particles.}$$

D)
$$v_{RMS} = \sqrt{3\frac{kT}{m}} = \sqrt{3\frac{1.38 \times 10^{-23} \times 294\,K}{M/N_A}} =$$
$$\sqrt{3\frac{(1.38 \times 10^{-23}\,J/K) \times (294\,K) \times (6.022 \times 10^{23}\frac{particles}{mol})}{28 \times 10^{-3} kg/mol}} = 511.64\,m/s.$$

95

Exercise A1.4

Consider the equation for the rate of arrival, $R = \rho\langle v\rangle/4$, show that indeed it equals $R = \dfrac{p}{T^{1/2}M^{1/2}}\sqrt{\dfrac{N_A}{2\pi k}}$, and evaluate the constant $\sqrt{N_A/2\pi k}$ to obtain a simplified expression in units of mbar, Kelvin degrees and molecular units for mass.

Solution

Since R=83.14 mol^{-1} K^{-1} = N$_A$k, k=1.38×10^{-23} J/K, M = N$_A$m.

$$r = \frac{1}{4}n\langle v\rangle = \frac{1}{4}n\left(\frac{8RT}{\pi m N_A}\right)^{1/2} = \frac{1}{4}\frac{p}{kT}\left(\frac{8RT}{\pi m N_A}\right)^{1/2} = \frac{1}{4}\frac{p}{kT}\left(\frac{8kN_A T}{\pi m N_A}\right)^{1/2}$$

$$= p\left(\frac{1}{kT\pi 2m}\right)^{1/2} = \frac{p}{T^{1/2}m^{1/2}}\left(\frac{1}{2\pi k}\right)^{1/2} = \frac{pN_A^{1/2}}{T^{1/2}M^{1/2}}\left(\frac{1}{2\pi k}\right)^{1/2}$$

$$= \frac{p}{T^{1/2}M^{1/2}}\left(\frac{N_A}{2\pi k}\right)^{1/2} = \frac{p}{T^{1/2}M^{1/2}}\left(6.9428\times10^{45}\right)^{1/2}$$

$$= \frac{P}{T^{1/2}M^{1/2}}8.3323\times10^{22}.$$

Now change units from Pa to mbar, g to kg and m to cm.

$$r = \frac{p}{T^{1/2}M^{1/2}}\left(\frac{100Pa}{1mbar}\right)\frac{\left(\frac{1000g}{1kg}\right)\left(\frac{100cm}{1m}\right)}{\left(\frac{1\times10^4 cm^2}{1m^2}\right)}2.6349\times10^{19}\ \frac{1}{cm^2 s}$$

$$= \frac{p}{T^{1/2}M^{1/2}}2.6349\times10^{22}\ \frac{1}{cm^2 s}.$$

This result yields particles per s and per cm^2, using p in mbar, T in K and M in g/mol.

Pressure measuring devices

Pressures can be measured with respect to an absolute vacuum or with respect to the atmospheric pressure. For instance, vehicle tire pressures are measured with respect to the atmospheric pressure; such pressure is known as *gauge* pressure; thus the absolute pressure of any system is the gauge pressure of the system plus atmospheric pressure. UHV systems always use the absolute pressure. Common vacuum classification is *Rough vacuum* for pressures

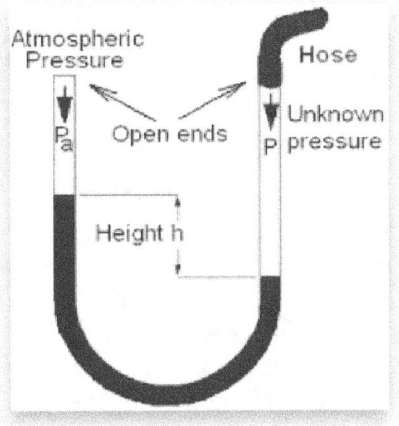

from atmospheric levels (1000 mbar) down to 0.01 mbar, *Process vacuum* from 10^{-2} mbar to 10^{-5} mbar, *High vacuum* from 10^{-5} mbar to 10^{-9} mbar, and UHV for pressures $< 10^{-9}$ mbar.

One of the first methods to measure pressure was by comparison to the atmospheric pressure using a "U" tube. If the tube contains a liquid of density ρ, the atmospheric pressure will push down through the left end of the open pipe, and an unknown fluid will push through the hose into the right open end of the pipe. The difference in pressures will push the liquid into the left or the right arms facilitating the measurement of the difference in column heights, difference which can be used to quantify the difference in pressure with respect to the atmospheric pressure. The two opposing forces are $p \times$ area and $p_a \times$ area, and their difference equals the weight of the displaced liquid, i.e.

$$(p - p_a) \times area = W = mg = \rho \times volume \times g = r \times area \times height \times g \Rightarrow$$
$$p = p_a + \rho hg,$$

where ρ is the density of the liquid and g is the acceleration of gravity 9.8 m/s^2. Customarily, the liquid used is mercury and the pressures determined this way are listed in mmHg or torr.

Perhaps the most common pressure meter is the *Bourdon tube* shown in the figure. Bourdon tubes have an inlet pipe closed with diaphragm that can be opened by gas pressure. Such diaphragm is

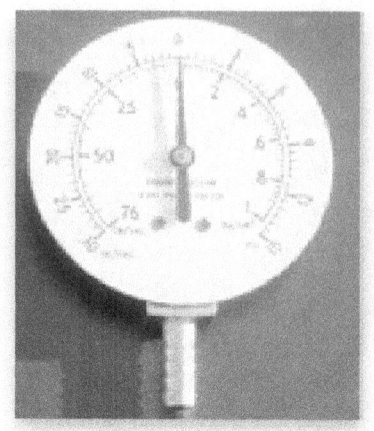

connected to a coil which rotates when the gas pressure opens it, moving a needle which indicates the pressure of the gas being measured. The range of operation is from 762 to 11201.4 mmHg with respect to atmospheric pressure, not absolute.

Another instrument is the *Thermocouple gauge tube*. As a gas increases in density, its ability to conduct heat also increases, since density is related to the pressure, the change of heat conductivity can be used to determine the pressure. These gauges use a thermocouple to measure the temperature of a wire which is dependent on the rate at which the filament loses heat to the surrounding gas, and therefore on the thermal conductivity. The 531 Varian tube, for instance, has a listed range of operation of 0.001 to 2 torr (www.varianinc.com).

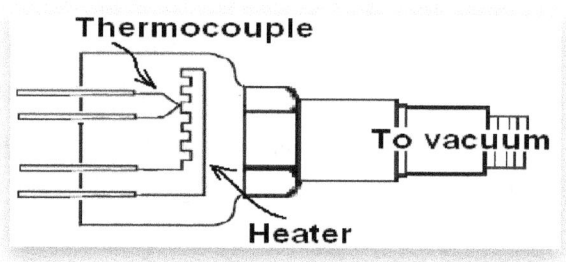

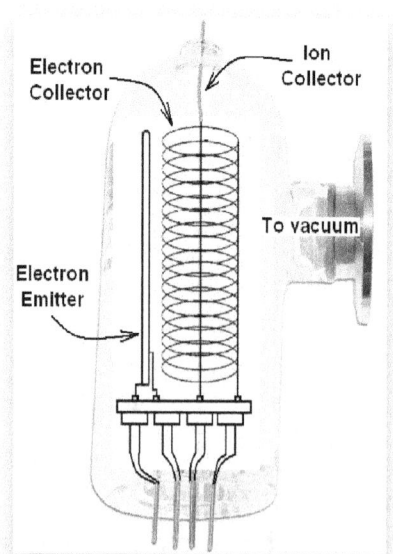

In the *Bayard-Alpert gauge* electrons emitted from a filament collide with gaseous molecules producing ionization, the number of ions is proportional to the gas density multiplied by the electron current emitted from the filament, the pressure is estimated by

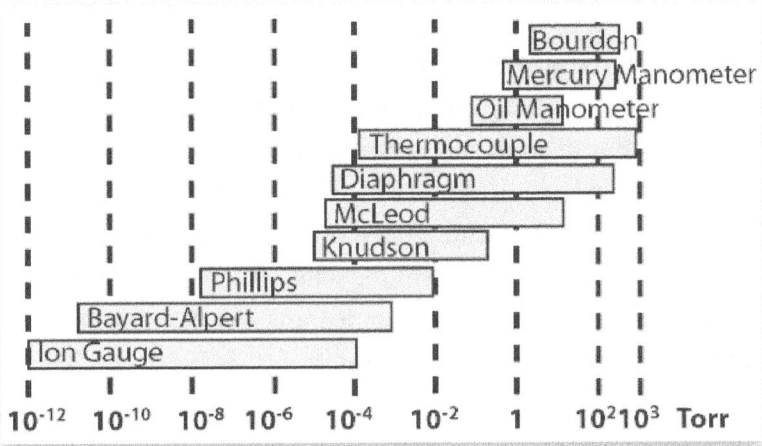

measuring the ion current. The picture shows the diagram of a Bayard Alpert ionization gauge which have ranges of operation of 10^{-3} to 2×10^{-11} torr.

Other methods exist for measuring low pressures; the chart lists different devices according to their range of operating pressures.

Pumps

Turbomolecular pumps take advantage of the large mean free path that molecules have at $p < 10^{-3}$ Torr, and use rotors to kick molecules out of the UHV chamber. Since the mean free path is larger than the space between rotor and stator blades, particles collide primarily with the rotor. The diagram shows the operation of a turbomolecular pump. The turbomolecular pump was invented in 1958 by Becker based on pumps of Holweck (1923) and Siegbahn (1944).

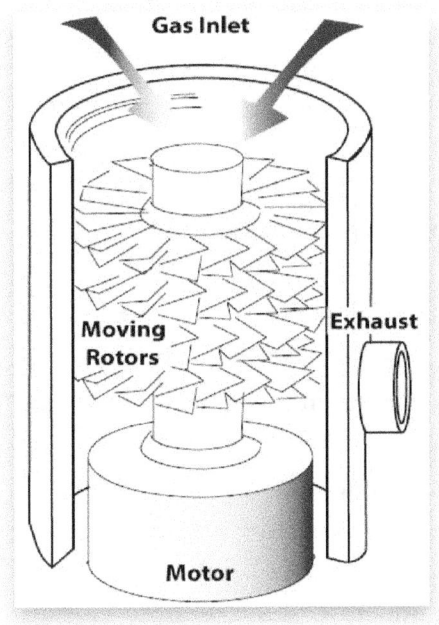

Ion pumps eliminate molecules by ionizing them first, and then using strong electrical potentials (typically 3 kV to 7 kV) to accelerate them into an electrode; they achieve pressures as low as 10^{-11} mbar. The pump is composed of a series of "Penning traps" (metal containers) where a plasma of electrons is created to ionize gas atoms. Once ionized, the gas atoms are accelerated onto a chemically active cathode of titanium, where the gas molecules will be buried or will adhere to Ti molecules and sputter onto the walls of the pump. The ion pump was invented in the 1950's by Helmer, Jepsen and Rutherford.

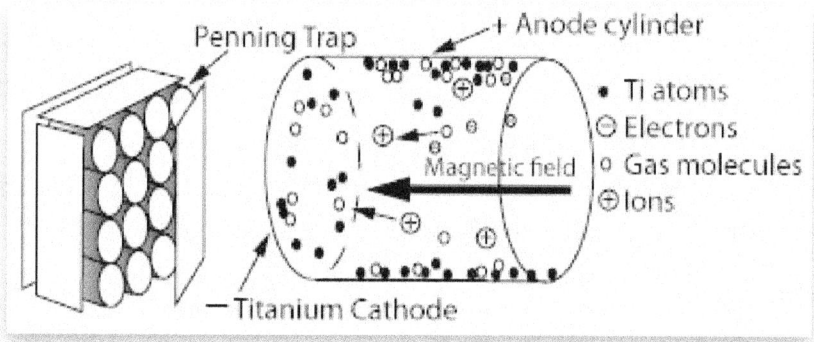

Diffusion pumps create a high speed jet of vapor to push gas molecules out into an exhaust. The vapor jet is Hg, Si oil or polyphenyl ethers. The working range of pressures is 10^{-10} to 10^{-2} mbar. Diffusion pumps were created by Gaede in 1915, and in the beginning it was thought that the working principle was diffusion, it is now known that it is momentum transfer what propels gas molecules out of the chamber.

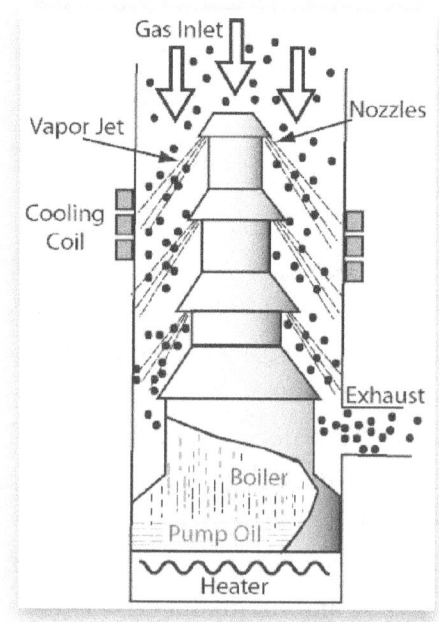

Rotary Vane Pumps operate with an eccentric rotor that admits gas from the chamber, compresses it during rotation and expels at a higher pressure into another hose. The range of operation is between 10^{-2} to 10^{-3} Torr. These pumps are mostly used as "roughing pumps".

Diaphragm Pumps use a flexible diaphragm on one end of a volume (bottom in figure), while two ends are covered with two spring-loaded valves (top). As the membrane flexes in or out of the volume, the

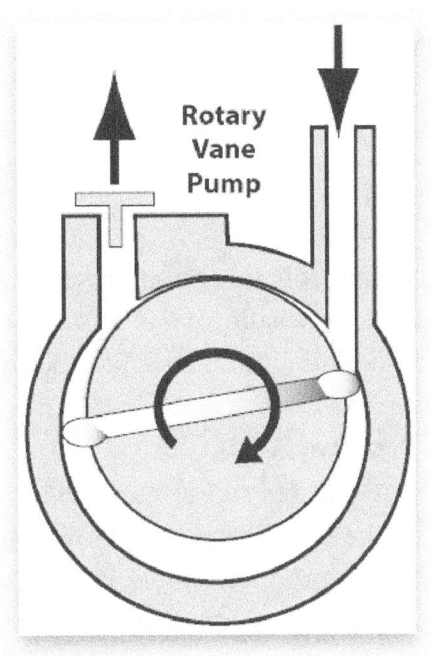

valves open one at a time to let gas into the chamber and then out of the chamber. A motor shaft rapidly flexes the diaphragm, causing gas transfer in one valve and out the other. The range of operation is between 1 to 10 Torr.

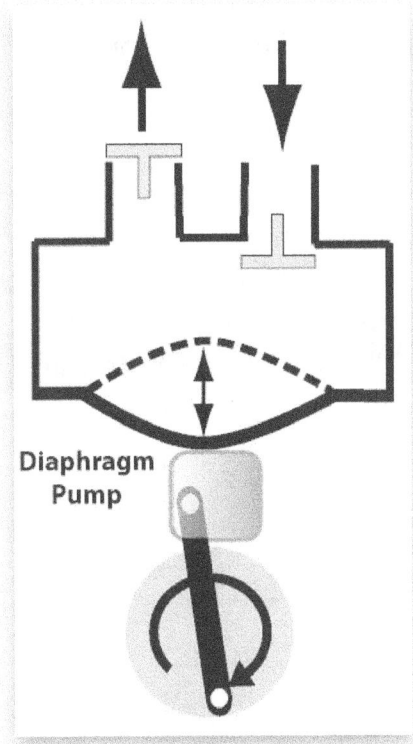

Cryogenic pumps trap gas molecules by condensing them on a cold surface. They are effective only on some gases depending on the range of temperatures of the cryo-pump. Cryo-pumps have wide pressure ranges of operation, from 10^{-3} torr to UHV values.

Problems

Problem A1.1

A UHV chamber contains *monatomic* nitrogen at room temperature and a pressure of 10^{-9} Torr. Determine the number of N particles that hit the sample surface per unit time per unit area.

Problem A1.2

A) With respect to the previous problem, assume nitrogen adheres to the sample surface. How long would it take for nitrogen at 10^{-9} torr to produce a monolayer coverage? Assume that a monolayer is about 1×10^{15} atoms/cm^2, and that the structure of the surface does not play a major role; although in practice it actually does.

B) Compare for the time needed when the pressure is 10^{-6} Torr and discuss whether UHV is needed in XPS or AES.

Problem A1.3

With respect to the previous problem, calculate the mean free path of *monatomic* nitrogen atoms in the UHV chamber for pressures of 1 torr, 10^{-5} torr and 10^{-9} torr, and sketch the behavior of the mean free path versus pressure in the range of 1×10^{-6} to 1×10^{-5} torr at room temperature (25 °C),

Appendix 2: Atomic Data

Table A2.1: Common XRF lines.

	Line	λ (nm)		Line	λ (nm)		Line	λ (nm)
Li	Kα	22.8	Fe	Kα_1	0.1936	In	Lα_1	0.377
Be	Kα	11.4	Co	Kα_1	0.1789	Sn	Lα_1	0.36
B	Kα	6.76	Ni	Kα_1	0.1658	Sb	Lα_1	0.344
C	Kα	4.47	Cu	Kα_1	0.1541	Te	Lα_1	0.329
N	Kα	3.16	Zn	Kα_1	0.1435	I	Lα_1	0.315
O	Kα	2.362	Ga	Kα_1	0.134	Xe	Lα_1	0.302
F	K$\alpha_{1,2}$	1.832	Ge	Kα_1	0.1254	Cs	Lα_1	0.289
Ne	K$\alpha_{1,2}$	1.461	As	Kα_1	0.1176	Ba	Lα_1	0.278
Na	K$\alpha_{1,2}$	1.191	Se	Kα_1	0.1105	La	Lα_1	0.267
Mg	K$\alpha_{1,2}$	0.989	Br	Kα_1	0.104	Ce	Lα_1	0.256
Al	K$\alpha_{1,2}$	0.834	Kr	Kα_1	0.098	Pr	Lα_1	0.246
Si	K$\alpha_{1,2}$	0.7126	Rb	Kα_1	0.0926	Nd	Lα_1	0.237
P	K$\alpha_{1,2}$	0.6158	Sr	Kα_1	0.0875	Pm	Lα_1	0.228
S	K$\alpha_{1,2}$	0.5373	Y	Kα_1	0.0829	Sm	Lα_1	0.22
Cl	K$\alpha_{1,2}$	0.4729	Zr	Kα_1	0.0786	Eu	Lα_1	0.212
Ar	K$\alpha_{1,2}$	0.4193	Nb	Kα_1	0.0746	Gd	Lα_1	0.205
K	K$\alpha_{1,2}$	0.3742	Mo	Kα_1	0.0709	Tb	Lα_1	0.198
Ca	K$\alpha_{1,2}$	0.3359	Tc	Kα_1	0.0675	Dy	Lα_1	0.191
Sc	K$\alpha_{1,2}$	0.3032	Ru	Kα_1	0.0643	Ho	Lα_1	0.185
Ti	K$\alpha_{1,2}$	0.2749	Rh	Kα_1	0.0614	Er	Lα_1	0.178
V	Kα_1	0.2504	Pd	Kα_1	0.0586	Tm	Lα_1	0.173
Cr	Kα_1	0.229	Ag	Kα_1	0.056	Yb	Lα_1	0.167
Mn	Kα_1	0.2102	Cd	Kα_1	0.0536	Lu	Lα_1	0.162

	Line	λ (nm)		Line	λ (nm)		Line	λ (nm)
Hf	Lα₁	0.157	Bi	Lα₁	0.114	Pu	Lα₁	0.087
Ta	Lα₁	0.152	Po	Lα₁	0.111	Am	Lα₁	0.085
W	Lα₁	0.148	At	Lα₁	0.109	Cm	Lα₁	0.083
Re	Lα₁	0.143	Rn	Lα₁	0.106	Bk	Lα₁	0.081
Os	Lα₁	0.139	Fr	Lα₁	0.103	Cf	Lα₁	0.079
Ir	Lα₁	0.135	Ra	Lα₁	0.101	Es	Lα₁	0.077
Pt	Lα₁	0.131	Ac	Lα₁	0.098	Fm	Lα₁	0.076
Au	Lα₁	0.128	Th	Lα₁	0.096	Md	Lα₁	0.074
Hg	Lα₁	0.124	Pa	Lα₁	0.093	No	Lα₁	0.072
Tl	Lα₁	0.121	U	Lα₁	0.091			
Pb	Lα₁	0.118	Np	Lα₁	0.089			

Table A2.2: Atomic energy levels.

All energies are in eV, adapted from: http://www.chembio. uoguelph.ca/educmat/atomdata/bindener/elecbind.htm.

	H	He	Li	Be	B	C
1s	13.5981	24.588	57.875	114.34	192.3	288.23
2s			5.3917	9.32263	12.93	16.59
2p					8.298	11.26
	N	O	F	Ne	Na	Mg
1s	403.78	538.25	692.45	869.5	1074.4	1307.3
2s	20.33	28.72	37.213	47.74	66.02	92.18
2p	14.534	13.618	17.423	21.57	33.52	53.7733
3s					5.13908	7.64624
	K	Ca	Sc	Ti	V	Cr
1s	3610.27	4041.1	4494.2	4969.9	5469.6	5994.2
2s	380.8	441.08	502.7	566.78	632.94	701.7

2p	297.25	350.22	405.183	461.093	520.607	583.507
3s	36.95	46.52	54.9	63.94	71.94	80.5
3p	19.375	28.08	32.6417	37.8533	42.9	48.6067
3d			8.06667	8.1	8.13333	8.34333
4s	4.34066	6.1132	6.56144	6.8282	6.7463	6.76664
	Mn	Fe	Co	Ni	Cu	Zn
1s	6542	7117	7713.4	8338	8983.9	9662
2s	773.28	851.06	930.96	1015	1102.9	1198
2p	648.033	717.43	789.21	865.8	944.78	1032
3s	88.44	97.92	106.58	117.4	125.42	141.2
3p	53.22	59.16	65.44	74.03	80.88	91.72
3d	9.1067	9.34	8.9867	10.21	10.62	12.28
4s	7.43402	7.9024	7.881	7.64	7.7264	9.394
	Al	Si	P	S	Cl	Ar
1s	1564.1	1844	2148.3	2476	2829.1	3206
2s	121.46	154	191.4	232.1	276.6	324.2
2p	76.753	103.7	135.12	168.1	206.35	247.7
3s	10.62	13.46	16.15	20.2	24.54	29.24
3p	5.9858	8.152	10.487	10.36	12.968	15.76
	Rb	Sr	Y	Zr	Nb	Mo
1s	15201.7	16106.6	17040	18000.6	18988.6	20003.5
2s	2068.07	2068.07	2374.8	2536.07	2702.13	2871.67
2p	1827.16	1965.11	2107.6	2254.96	2406.38	2561.44
3s	325.55	360.94	396.05	433.98	471.48	510.5
3p	244.733	275.087	306.33	338.613	369.007	404.147
3d	114.1	137.048	160.34	184.456	208.644	234.044
4s	32.4	40.66	47.95	55.04	61.12	67.56
4p	16.5167	22.7733	28.625	32.9667	37.3133	41.46
4d			6.38	8.61	7.17	8.56

		Tc	Ru	Rh	Ag	Cd
5s	4.17713	5.69484	6.217	6.6339	6.75885	7.09243
1s		21048	22121.2	23223.9	25518	26714
2s		3048.33	3230	3417.93	3812	4022.3
2p		2721.78	2886.93	3057.27	3415	3605.2
3s		550.333	590.82	633.06	723.5	774.88
3p		437.983	473.513	509.607	587.5	632.48
3d		260.35	286.608	314.168	374.9	410.86
4s		74.2667	80.16	86.28	101.4	112.4
4p		44.775	47.44	53.6917	64.49	71.073
4d		8.6	8.5	9.56	10.4	13.4
5s		7.28	7.3605	7.4589	7.576	8.9937

	Cs	Ba	La	Hf	Ta	W
1s	35985	37441	38925	65350.8	67416.4	69525
2s	5714.3	5988.8	6266.3	11270.7	11681.5	12099.8
2p	5127.7	5372.5	5618.7	9953.6	10299.4	10652.5
3s	1218.3	1292.7	1362.5	2601.93	2709.33	2820.73
3p	1021.8	1087.7	1151.8	2195.13	2286.6	2379.64
3d	731.85	787.01	840.11	1684.72	1759.09	1834.84
4s	232.15	253.5	272.2	538.05	565.48	595.36
4p	165.67	183.87	198.85	400.547	424.053	447.44
4d	78.21	91.372	101.45	217.516	233.44	250.256
4f				16.8857	25.1086	34.3971
5s	23.85	30.1	34.15	65	70.98	76.94
5p	12.267		16.7	33.7083	38.19	40.8783
5d			5.75	7	8.3	9
6s	3.8939	5.2117	5.577	6.82507	7.89	7.98

	Re	Os	Ir	Pt	Au	Hg
1s	71676.4	73870.8	76111	78395	80725	83102

2s	12526.7	12968	13419	13880	14353	14839
2p	11009.8	11375.1	11752	12133	12524	12926
3s	2933.47	3050.33	3174.1	3297.7	3427	3563.4
3p	2473.38	2570.24	2671.1	2773.7	2880	2992.5
3d	1910.67	1989.72	2071.8	2155.1	2241	2332.6
4s	625.88	656.96	691.26	724.5	761.4	802.76
4p	470.513	496.22	522.92	549.79	578.9	610.15
4d	266.452	282.764	302.59	321.72	342.6	368.11
4f	43.1143	51.5	62.697	73.183	85.92	102.05
5s	86.72	87.1	96.35	103.1	109.4	123.52
5p	41.4667	50.1067	54.547	57.6	62.61	70.127
5d	9.6	9.6	9.6	9.6	11.66	12.8
6s	7.88	8.4382	8.967	9	9.226	10.438
	Pd	Ga	Ge	As	Se	Br
1s	24355	10367	11105	11870	12660	13478
2s	3611	1302.1	1415.9	1531	1656.6	1787.7
2p	3232.4	1128.1	1230	1339.1	1452.3	1571.3
3s	676.46	162.04	183.25	208	234.25	261.75
3p	546.75	107.08	126.08	146.65	167.33	190.05
3d	342.9	21.14	31.82	45.555	59.575	75.17
4s	92.78	11	14.3	17	20.15	23.8
4p	56.887	5.9993	7.9	9.8152	9.7524	11.814
4d	8.5141					
	In	Sn	Sb	Te	I	Xe
1s	27943	29203	30494	31818	33174	34563
2s	4242	4469	4701.3	4943.2	5193	5455
2p	3804	4009	4219.6	4436.8	4660	4891
3s	830.28	888.2	948.28	1011.8	1078	1148
3p	681.25	732.5	785.51	841.47	899.9	959.8

3d	450.33	492	535.53	581.82	630.6	681
4s	126.14	140.6	156.44	174	193.2	214.6
4p	81.367	92.37	103.01	115.9	132.4	153.5
4d	20.416	28.12	36.148	47.168	57.46	68.15
4f						
5s	10	12	15	17.84	20.61	23.39
5p	5.7864	7.344	8.64	9.0096	10.45	12.56
	Tl	Pb	Bi	Po	At	Rn
1s	85530	88005	90526	93102	95727	98401
2s	15347	15861	16388	16934	17487	18049
2p	13338	13757	14183	14622	15068	15521
3s	3706.1	3853	4001.7	4154.6	4316.2	4480
3p	3111.1	3231	3352.7	3482.6	3613.5	3724
3d	2429.2	2527	2625.6	2729	2833.8	2952
4s	847.28	894.2	940.8	992.27	1041.4	1095
4p	647.6	685.1	723.69	760.92	799.68	837.8
4d	394.89	422.6	452.16	484.11	515.2	546.6
4f	120.99	140.6	161.95	185.46	208.51	232.2
5s	136.15	149.4	161.87	179.25	194.7	210.8
5p	82.627	92.61	103.56	117.03	128.63	140.4
5d	14.988	20.9	27.76	34.16	42.2	49.16
6s	8	10	12	15	19	24
6p	6.1083	7.417	7.289	8.4167	9.8667	11.83
	Ce	Pr	Nd	Pm	Sm	Eu
1s	40443	41991	43569	45184	46834	48519
2s	6548.8	6834.8	7126	7427.9	7736.8	8052
2p	5870.3	6123	6379.1	6643.8	6914.7	7190.3
3s	1435.4	1510.3	1576.9	1650.8	1724.5	1801.7
3p	1214.2	1274.7	1334.6	1396.7	1461.2	1526.2

3d	891.1	940.02	988.69	1040	1091.7	1142.4
4s	290.05	305.25	317.85	334.47	347.35	361.85
4p	212.95	224.25	233	245.18	256.57	266.93
4d	110.6	115.01	119.89	110.6	130.25	133.45
4f	6	6	6	6	6	6
5s	37.55	38.45	39.5	39	39.95	36.9
5p	20.15	21.908	21.792	21.856	21.775	22.583
6s	5.5387	5.464	5.525	5.55	5.6437	5.6704

	Tb	Dy	Ho	Er	Tm	Yb
1s	51996	53789	55618	57486	59390	61332
2s	8708	9045.8	9394.2	9751.3	10116	10486
2p	7759.9	8053.6	8353.1	8660	8971	9288.5
3s	1966.7	2047.2	2129.9	2214.4	2307.5	2399.1
3p	1663.6	1732	1803.2	1877.4	1953.3	2026.1
3d	1255.3	1310.9	1368.4	1427.8	1487.6	1548.3
4s	397.45	416.4	435.1	450.3	471.1	486.6
4p	294.78	307.6	321.03	336.12	351.85	361.28
4d	148.2	155.05	162.31	171.27	179.71	189.09
4f	6	6	6	6	7	7.4286
5s	43.5	55.95	51.6	56.4	53.85	54.3
5p	25.175	25.908	24.65	27.892	30.992	26.45
6s	5.8639	5.9389	6.0216	6.1078	6.1843	6.2542

	Th	Pa	U	Np	Pu	Am
1s	109650	112599	115603	118673	121805	125004
2s	20472	21108	21757	22427	23098	23789
2p	17431	17925	18427	18941	19460	19986
3s	5182.3	5366.9	5548	5731.4	5932.9	6126.6
3p	4308.2	4450.7	4597.4	4746.2	4884.8	5033.9
3d	3396.5	3509.5	3623.8	3740.5	3857.5	3972.5

4s	1330.5	1388.1	1442.5	1501.6	1560.3	1619.8
4p	1034.6	1080.3	1121.3	1167.4	1206.7	1247.5
4d	691.58	724.19	755.1	789.62	821.57	851.75
4f	338.79	365.1	384.78	409.72	434.96	456.19
5s	291.85	310.4	324.5	342.9	353.27	359.67
5p	198.13	213.77	218.13	231.64	236.53	245.64
5d	90.064	95.447	99.7	106.04	111.19	112.41
5f		6	6	6	6	6
6s	50	48.35	52.02	52	50.8	52.2
6p	27.667	26.017	26.66	27.033	25.956	35.789
6d	6	6	6.1	6		
7s	6.08	5.89	6.1941	6.2657	6.06	5.993
	Bk	Cf	Es	Fm	Md	No
1s	131573	134952	138446	142010	145658	149379
2s	25267	26065	26843	27637	28462	29304
2p	21089	21673	22247	22833	23425	24029
3s	6551.8	6759	6981.2	7208.8	7443.3	7680
3p	5365.3	5526.5	5693.4	5863.7	6037	6210.3
3d	4223.5	4346.6	4473	4602.7	4732.3	4860.8
4s	1744.1	1801.3	1869.7	1941.4	2011.4	2083.1
4p	1344.5	1392.2	1442.1	1495.5	1545.7	1596.5
4d	922.36	954.85	989.5	1028.9	1073	1099.6
4f	508.09	532.64	560.11	590.02	626.95	640.7
5s	401.33	418.2	435.53	456.13	478.7	489.4
5p	273.93	279.92	291.3	304.83	319.98	326.07
5d	126.21	131.1	138.37	146.74	152.1	155.12
5f	12	9	9	15	12.286	12.286
6s	59.35	59.55	61.7	65.85	67.05	68.25
6p	30.5	28.767	29.617	33.35	31.417	31.667

7s	6.23	6.3	6.42	6.5	6.58	6.65
	Cm	Ac	Lw	Fr	Ra	
1s	128230	106756	153181	101133	103919	
2s	24494	19843	30165.5	18636.8	18637	
2p	20541	16942	24639.7	15986.8	16458	
3s	6315.2	5001.3	7918.55	4648.85	4822	
3p	5185.9	4162.5	6347	3882.18	4023.1	
3d	4092.3	3280.9	4989.5	3051.88	3164.3	
4s	1665.2	1270.8	2158.7	1151.67	1210.3	
4p	1283.4	977.04	1650.2	939.956	939.96	
4d	886.4	656.34	1144.1	583.967	618.05	
4f	481.14	314.34	675.01	259.014	291.12	
5s	387.23	273.75	509.2	230.95	253.93	
5p	261.44	186.42	344.95	155.517	170.56	
5d	120.3	82.47	166.95	59.98	69.28	
5f	11		18.286			
6s	58.45	45	75.3	33	40	
6p	29.456	24.33333	36.183	15.6667	21	
6d	5	5.7	4			
7s	6.02	5.17	7	4	5.2789	
	Kr	Gd	Lu			
1s	14328	50239	6314			
2s	1924.7	8375.6	10870			
2p	1696	7472	9612.3			
3s	293.13	1881.9	2493.8			
3p	217.85	1593.2	2106.2			
3d	93.722	1199.1	1611.4			
4s	27.51	377.4	508.85			
4p	14	278.47	378.53			

4d		141.7	201.06			
4f		6	12.429			
5s		40.3	58.15			
5p		22.483	29.917			
5d		6	6.6			
6s		6.15	5.4258			

Table A2.3:

Layer thickness (in μm) from which 90% of the fluorescence radiation originates in graphite, glass, iron or lead, respectively, when excited by x ray radiation. For instance when glass is irradiated with x rays, the tungsten impurities it contains, will produce fluorescence in a layer of thickness of 429 μm. The same type of material, tungsten, in a lead matrix will fluoresce within the top 22.4 μm from the surface irradiated. Adapted from the manuscript of Schlotz and Uhlig.

	Line	Graphite	Glass	Iron	Lead
U	Lα1	28000	1735	154	22.4
Pb	Lβ1	22200	1398	125	63.9
Hg	Lα1	10750	709	65.6	34.9
W	Lα1	6289	429	40.9	22.4
Ce	Lβ1	1484	113	96.1	6.72
Ba	Lα1	893	71.3	61.3	4.4
Sn	Lα1	399	44.8	30.2	3.34
Cd	Kα1	144600	8197	701	77.3
Mo	Kα1	60580	3600	314	36.7
Zr	Kα1	44130	2668	235	28.9
Sr	Kα1	31620	1947	173	24.6
Br	Kα1,2	18580	1183	106	55.1
As	Kβ1	17773	1132	102	53
Zn	Kα1,2	6861	466	44.1	24
Cu	Kα1,2	5512	380	36.4	20
Ni	Kα1,2	4394	307	29.8	16.6
Fe	Kα1,2	2720	196	164	11.1
Mn	Kα1,2	2110	155	131	9.01
Cr	Kα1,2	1619	122	104	7.23
Ti	Kα1,2	920	73.3	63	4.52
Ca	Kα1,2	495	54.3	36.5	3.41
K	Kα1,2	355	40.2	27.2	3.04
Cl	Kα1,2	172	20.9	14.3	2.19
S	Kα1,2	116	14.8	10.1	4.83
Si	Kα1,2	48.9	16.1	4.69	2.47
Al	Kα1,2	31.8	10.5	3.05	1.7
Mg	Kα1,2	20	7.08	1.92	1.13

Table A2.4.

Atomic radii as determined by Slater in 1964 according to the distance between atoms in covalent bonds; the radii of inert gases (and other elements) cannot be determined this way.

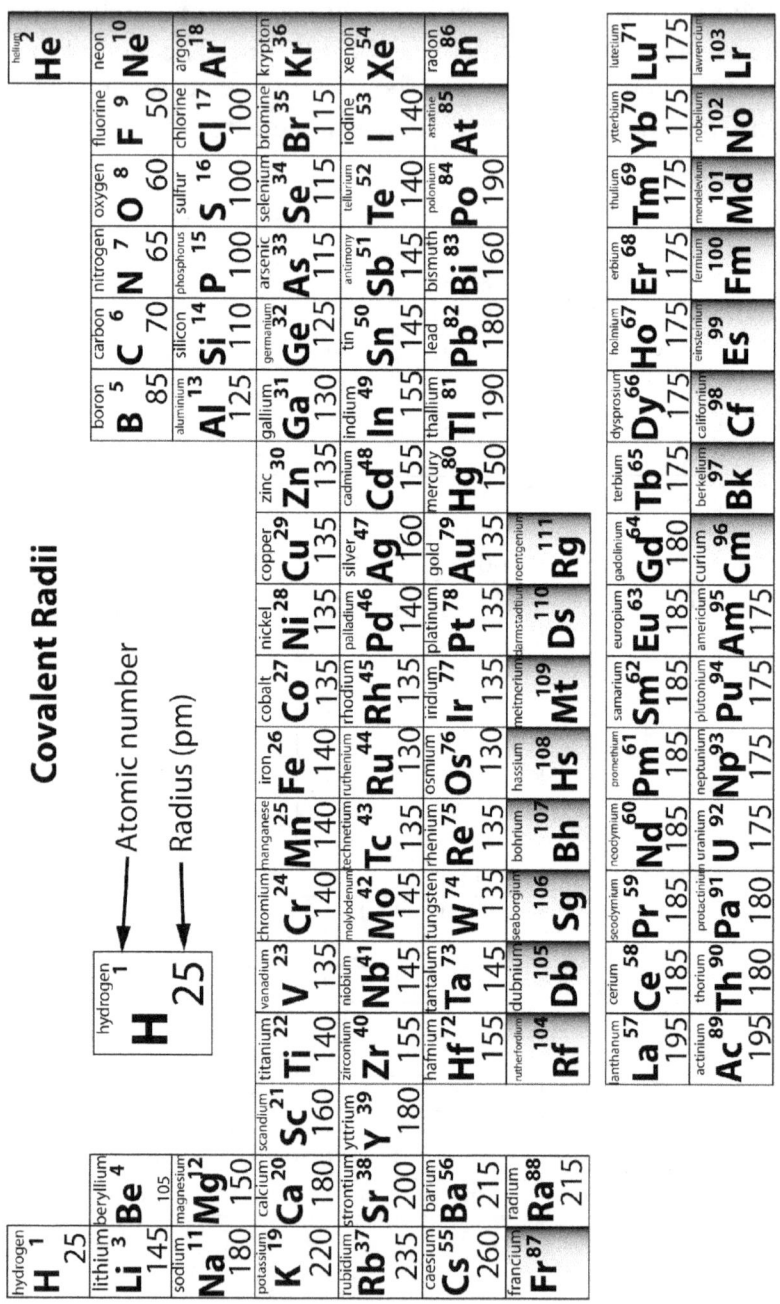

Appendix 3: Comparison of Techniques

Figure A3.1: Physical processes of the three main spectroscopic techniques.

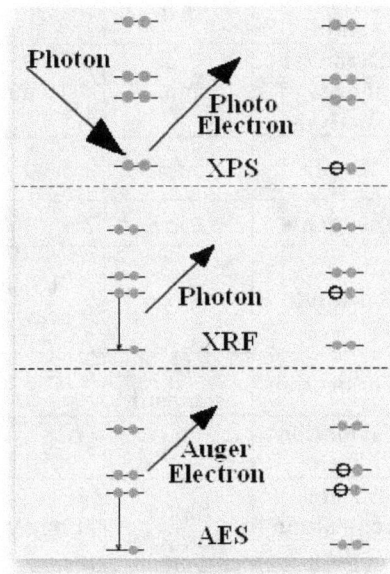

Figure A3.2: Depths of emitted radiation.

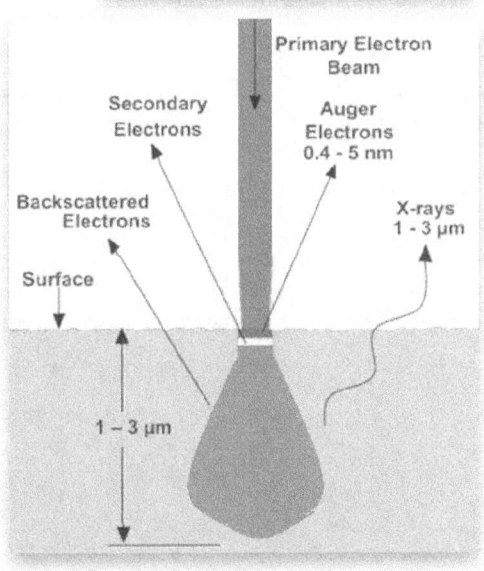

Figure A3.3: Comparison of properties of the main spectroscopic techniques.

	XRF	XPS	AES	SIMS
Primary beam	x rays	x rays	electrons	ions
Produced particles	x rays	electrons	electrons	ions
Types of samples	All	All	Conducting	All
Diameter of area of analysis	mm	10 mm - 5 mm	10 nm	100 nm
Depth	mm	1 - 5 nm	1 - 5 nm	0.1-1 nm
Elements	$Z \geq 3$	$Z \geq 3$	$Z \geq 3$	$Z \geq 3$
Sensitivity	ppt	0.1 % highest for heavy elements	0.1 % highest for light elements	1 ppm
Calibration	by standard	by standard	by standard	by standard
Oxidation state		Yes	Yes	Not
Data acquisition time	Faster than XPS	Longer than AES	Shorter than XPS	
Destructive	Not	Not	Electron beam can damage	Yes

116

Figure A3.4: Analysis depths of main surface techniques, adapted from Evans Analytical Group.

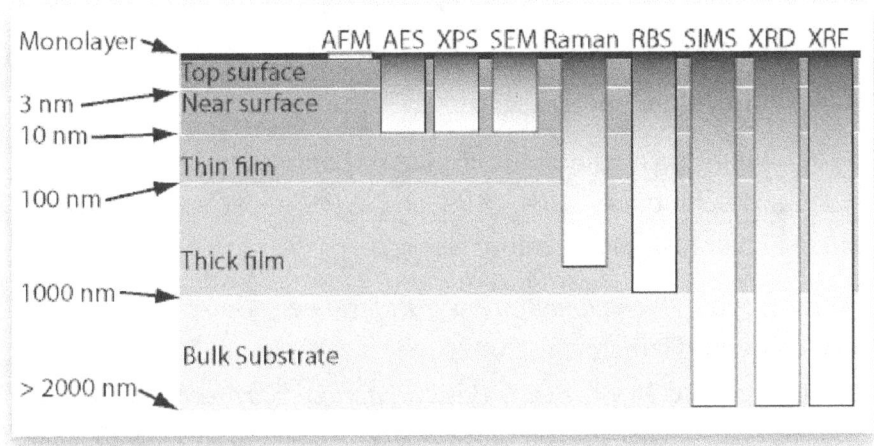

Figure A3.5: Detection capabilities of the main spectroscopic techniques.

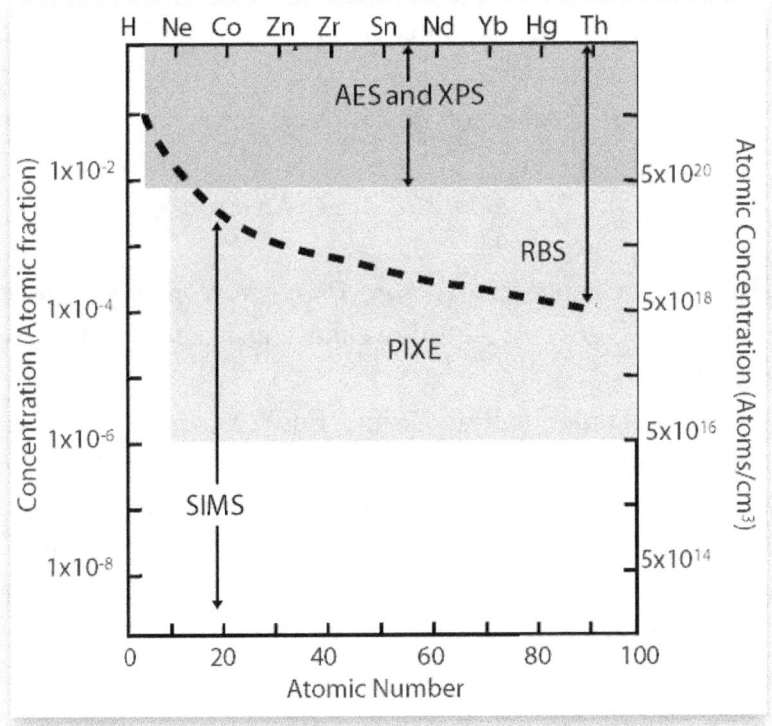

Bibliography

General sources on spectroscopy

The following books and reading material are specialized treatises on vacuum techniques, XRF, XPS and AES; most are at a more advanced level than these lecture notes.

- W. Boyes, *Instrumentation Reference Book*, 3rd Ed., Butterworth-Heinemann, 2002.
- D. Briggs and M. P. Seah, Eds., *Practical Surface Analysis by Auger and X-ray Photoelectron Spectroscopy*, Wiley, 1983.
- D. Briggs and M.P. Seah, *Practical Surface Analysis. Volume 1–Auger and X-ray Photoelectron Spectroscopy*, 2nd Ed., Wiley and Sons, 1990.
- CasaXPS, processing software for XPS, AES, SIMS, etc., http://www.casaxps.com/.
- T. A. Carlson, *Photoelectron and Auger Spectroscopy*, Plenum Press, 1975.
- C. L. Hedberg, Ed., *Handbook of Auger Electron Spectroscopy*, Physical Electronics, 1995.
- K. Janssens, Chapter 2: *X-Ray Photoelectron and Auger Electron Spectroscopy*, http://webhost.ua.ac.be/mitac4/micro_xpsaes.pdf.
- D. C. Konigsberger and R. Prins, Eds., *X ray Absorption: Principles, Applications, Techniques of EXAFS, SEXAFS and XANES*, in Chemical Analysis 92, John Wiley and Sons, 1988.
- J. F. Moulder, W. F. Stickle, P. E. Sobol and K. D. Bomben, *Handbook of X-ray Photoelectron Spectroscopy*, J. Chastain and R. C. King Jr., Eds., Physical Electronics, 1995.
- V. I. Nefedov, *X-ray Photoelectron Spectroscopy of Solid Surfaces*, English Edition, VSP BV, 1988.

- M. Newville, *Fundamentals of XAFS*, http://www.documbase. com/Fundamentals-of-XAFS.pdf, also found at xafs.org.
- J. J. Rehr and R. C. Albers, *Theoretical approaches to x-ray absorption fine structure*, Rev. Mod. Phys, 72, 621-982, 2000.
- M. S. Shackley, Ed., *X ray Fluorescence Spectrometry (XRF) in Geoarcheology*, Springer, 2011.
- J. Stoehr, *NEXAFS spectroscopy*, Springer, 1992.
- Surface Analysis Forum, www.uksaf.org.
- M. Thompson, M. D. Baker, A. Christie and J. F. Tyson, *Auger Electron Spectroscopy*, John Wiley and Sons, 1985.
- J. M. Walls, Ed., *Methods of Surface Analysis*, Cambridge University Press, 1989.
- J. F. Watts, *An Introduction to Surface Analysis by Electron Spectroscopy*, Oxford University Press, 1990.
- Xafs.org, a wiki community for x-ray absorption fine-structure (XAFS) and related spectroscopies.

Sources for Chapters 1 and 2

The books are general sources for properties of particles, radiation and their interactions. Hypherphysics is a comprehensive site for physics concepts, Richard's site provides a Bragg diffraction applet, and NIST's site provides a wealth of information including physical constants and atomic weights of elements.

- A. Beiser, *Concepts of Modern Physics*, McGraw Hill, 2002,
- C. R. Nave, Hyperphysics web site, Georgia State University, http://hyperphysics.phy-astr.gsu.edu/hbase/hph.html.
- K. Krane, *Modern Physics*, 3rd Ed., John Wiley & Sons, 2012.
- National Institute of Standards and Technology
  - *Reference on Constants, Units, and Uncertainty*, http://physics.nist.gov/cuu/Constants/index.html.
  - *Atomic weights of elements*, http://www.nist.gov/pml /data/comp.cfm.
- P. Tipler and R. Llewellyn, *Modern Physics*, 6th Ed., W. H. Freeman, 2012.

- G. A. Richard, *Bragg diffraction applet*, http://serc.carleton.edu /NAGTWorkshops/deepearth/activities/40414.html
- R. Serway and C. Moses, *Modern Physics*, 3rd Ed., Brooks Cole, 2004.

XRF Sources

In addition to the general sources, the sites listed here provide specific information of particular interest to XRF, such as x ray emission energies and fluorescent yields including an applet and a link to a request form for a free x ray data booklet.

- *Fluorescent yields*, Lawrence Berkeley Lab, http://xdb.lbl. gov/Section1/Sec_1-3.html.
- *X ray emission energies* (all from Lawrence Berkeley Lab.)
  - http://xdb.lbl.gov/Section1/Sec_1-2.html.
  - http://xdb.lbl.gov/Section1/Periodic_Table/X-ray_Elements.html.
  - Applet, http://henke.lbl.gov/optical_constants/pert_ form.html.
- *X ray data free booklet*, http://cxro.lbl.gov/x-ray-data-booklet.
- *The Decay Data Evaluation Project*, www.nucleide.org, has comprehensive information about atomic energy levels.

XPS Sources

In addition to the general sources, the sites listed here provide atomic binding energies, XPS spectra, electron mean free paths, among other data.

- Atomic binding energy
  - D. Thomas, *Table of atomic binding energies*, www. chembio.uoguelph.ca/educmat/atomdata/bindener/elecbi nd.htm.
  - Lawrence Berkeley Lab., http://xdb.lbl.gov/Section1/Se c_1-1.html.
  - Lawrence Berkeley Lab., http://xdb.lbl.gov/Section1/Se c_1-8.html.

- o WebElements, Tables of energy levels can be found in www.webelements.com/.
  - o www.nucleide.org.
- NIST
  - o *Basic atomic spectroscopic data*, http://www.nist.gov /pml/data/handbook/index.cfm.
  - o *Electron InelasticMean Free Path Database*, 2000, U.S. Secretary of Commerce, http://www.nist.gov/srd/nist71 .cfm.
- *XPS Spectra*, handbooks and reference material, www.xpsdata. com.
- *X Ray Photoelectron Spectroscopy Reference Pages*, www. xpsfitting.com/. This site contains information gained from decades of X-ray photoelectron spectroscopy (XPS) analyses of an enormous variety of samples analyzed at Surface Science Western Laboratories at the University of Western Ontario.

AES Sources

In addition to the general sources, the sites listed here provide atomic binding energies, XPS spectra, electron mean free paths, among other data.

- *Auger electron energies*
  - o Lawrence Berkeley Lab, http://xdb.lbl.gov/Section1 /Sec_1-4.html.
  - o K. D. Childs, et al. in C. L. Hedberg, Ed., *Handbook of Auger Electron Spectroscopy*, Physical Electronics, 1995. Provides an extensive table of Auger Peaks.
- *AES Training Tutorials*, Evans Analytical Group, http://www. eaglabs.com/mc/aes-auger-electron-energies.html.

References

- M. R. Alexander, G. E. Thompson, X. Zhou, G. Beamson and N. Fairley, *Quantification of oxide film thickness at the surface of aluminum using XPS*, Surf. Interface Anal. 34, 485, 2002.

- C. Altavilla and E. Ciliberto, Appl. Phys. A79, 309, 2004.

- X Ray Photoelectron Spectroscopy Reference Pages, *Aluminum oxide thickness measurement*, www.xpsfitting.com/2009/04/aluminum-oxide-thickness-measurement.html.

- R. Baranowski, A. Rybak, I. Baranowska, *Speciation Analysis of Elements in Soil Samples by XRF*, Polish Journal of Environmental Studies 11, 473, 2002.

- R. K. Brow and C. G. Pantano, *Compositionally Dependent Si 2p Binding Energy Shifts in Silicon Oxynitride Thin Films*, J. Am. Ceram. Soc. 69, 314, 1986.

- R. Schlotz and S. Uhlig, *Introduction to X Ray Fluorescence XRF*, Bruker AXS GmbH, Karlruhe, West Germany, 2006.

- B. V. Crist, http://en.wikipedia.org/wiki/File:XPS_PHYSICS.png.

- H. Guo, W. Maus-Friedrichs, and V. Kempter. Surf. Interf. Analysis 25, 390, 1997.

- R. Linke, M. Sehreiner, G. Demortier, M. Alram, H. Winter, *Chapter 13 The provenance of medieval silver coins: analysis with EDXRF, SEM/EDX and PIXE*, in *Non-Destructive Microanalysis of Cultural Heritage Materials*, Comprehensive Analytical Chemistry, 42, 605, 2004.

- Queen Mary University of London: www.chem.qmul.ac.uk/surfaces/scc/scat5_3.htm

- E. Pegg, Fig. 7 in *Peak Fitting in XPS*, CasaXPS, www.casaxps.com.

- G. D. Rusche, *XRF calibration Homework.docx*, University of Vermont, http://ebookbrowse.com/xrf-calibration-homework-docx-d33329821.
- T. A. Savas et al., *Large-area achromatic interferometric lithography for 100 nm period gratings and grids*, J. Vac. Sci. Tech. B14, 4167, 1996; T. A. Savas et al., *Achromatic interferometric lithography for 100-nm-period gratings and grids*, J. Vac. Sci. Tech. B13, 2732, 1995.
- M. L. Schattenburg et al., *x-ray/vuv transmission gratings for astrophysical and laboratory applications*, Phys. Scripta 41, 13, 1990.
- M. P. Seah and W.A. Dench, Surf. Interface. Anal. 1, 2, 1979.
- J.C. Slater, *Atomic Radii in Crystals*, J. Chemical Phys. 41, 3199, 1964.
- *Oxide Thickness Calculator in Excel*, http://sprocket.ssw.uwo.ca/xpsfiles/.
- B. R. Strohmeier, Surf. Interface Anal. 15, 51, 1990.

About the authors

Jorge López Gallardo is ρστ-Schumaker Professor of physics at the University of Texas at El Paso and has performed research at The Cyclotron Institute of Texas A&M University, The Niels Bohr Institute in Copenhagen, The Lawrence Berkeley Lab and
other institutions. Fellow of the American Physical Society (APS) and Corresponding Member of the Mexican Academy of Sciences, Dr. López received SHEP's 2011 Educator of the Year Award, MAES' 2010 MAEStro Award and the 2009 Hyer Research Award of the Texas Section of APS. He is the author of "Phase Transitions in Nuclear Matter" (World Scientific, 2000) and of hundreds of research and dissemination articles and three other books on scientific studies of electoral data.

Miguel Castro Colín is an expert in X-ray scattering. Currently at Bruker AXS GmbH in Karlsruhe, Germany, Dr. Castro has held research positions at the National Institute of Nuclear Research of Mexico, the Laser Plasma Physics Group at the University of
Oxford, the X-ray Scattering Group at The University of Houston and at the Max Planck Institute in Stuttgart Germany. He has performed spectroscopic and elastic scattering measurements in facilities in the United Kingdom (Rutherford Appleton Lab.), USA (Brookhaven National Lab., Argonne National Lab. and LCLS in Stanford), Germany (DESY in Hamburg and the ANKA Synchrotron in Karlsruhe) and France (ESRF).

www.ingramcontent.com/pod-product-compliance
Lightning Source LLC
Chambersburg PA
CBHW051320170526
45166CB00002B/616